AF885307

I Can Do Math Practice Problems GRADE 2

Volume 1: Units 1 - 4

This book belongs to:

____

I Can Do Math Practice Problems GRADE 2

Volume 1: Units 1 - 4

Dr. Christine Scafidi

Published by CKingEducation

CKingEducation

To contact CKingEducation or the authors about speaking, workshops, or ordering books in bulk, visit www.ckingeducation.com.

ISBN: 978-1-968264-00-0

Lead writer: Dr. Christine Scafidi
Editor: John Sasko
Series editor and book design: Christine King

Printed in the United States of America

Table of Contents

Track what you have done by checking off items.

Table of Contents

Track what you have done by checking off items.

Table of Contents

Track what you have done by checking off items.

Table of Contents

Track what you have done by checking off items.

Table of Contents

Track what you have done by checking off items.

Welcome to 2nd Grade!

Doing these practice problems will support you to better understand the skills and concepts you are learning.

Unit 1

Adding, Subtracting, and Working with Data

In this unit, we will solve math stories using graphs and diagrams. We will practice adding within 100 and subtracting within 20 to help us understand and solve different kinds of problems.

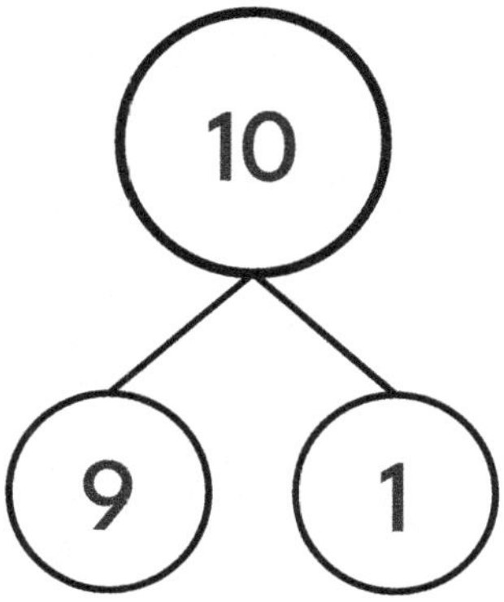

Number Bond

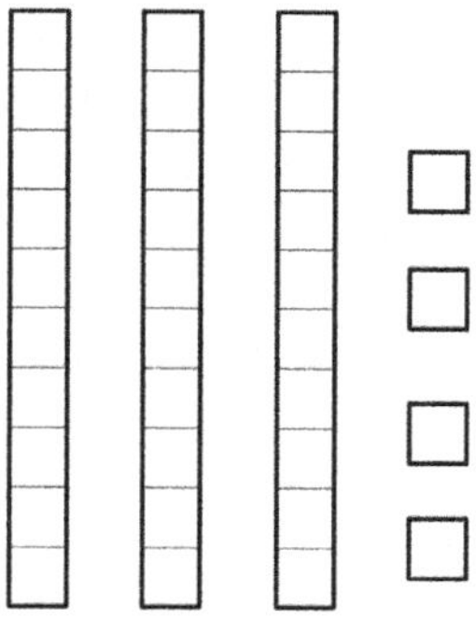

Base Ten Blocks

Cube Towers

Art Supplies	Number of Each
Pencils	15
Markers	10
Crayons	20

Data Table

How many fewer markers than crayons are listed on the Art Supplies data table?

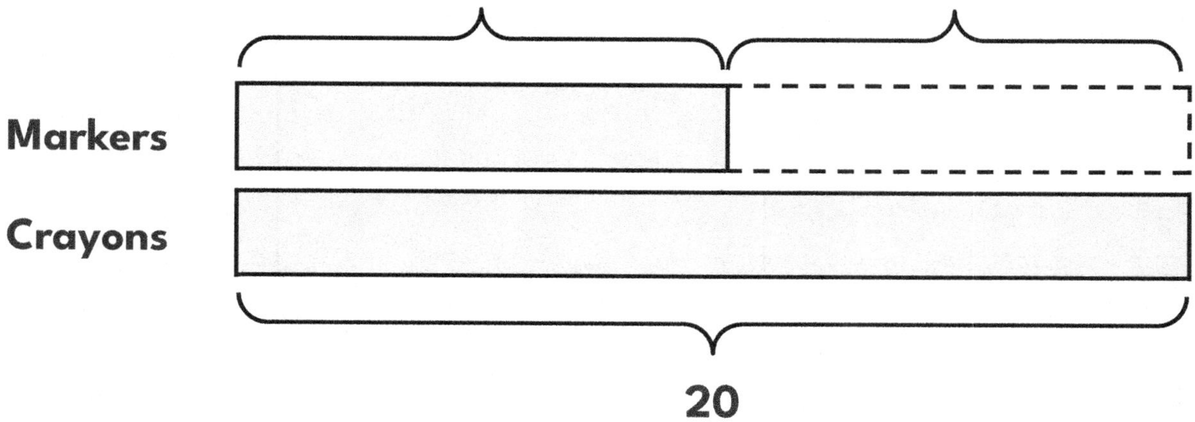

Unit Vocabulary

Adding, Subtracting, and Working with Data

Use this space to visualize the math vocabulary for this unit.

Word or Phrase	Example or Attributes	Visual Reminder

Unit Models & Strategies

Adding, Subtracting, and Working with Data

Use this space to visualize the math models and strategies for this unit.

Model or Strategy	This is a...	It is used to...

Name: ____________________ Date: ____________

Unit 1 Practice 1: I can add and subtract within 10.

1) The number bond shows one way you can break apart 10. Use each equation to break apart 10 in a different way.

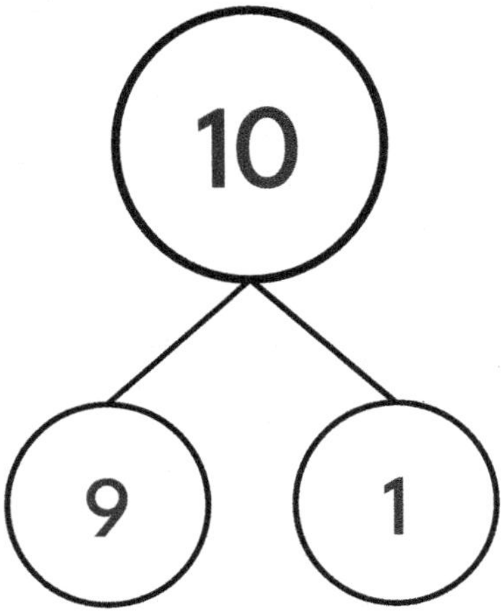

$10 = \boxed{9} + \boxed{1}$ $\quad$ $10 = \square + \square$

$10 = \square + \square$ $\quad$ $10 = \square + \square$

$10 = \square + \square$ $\quad$ $10 = \square + \square$

1.OA.6

2) Use the cube tower to help you find the difference.

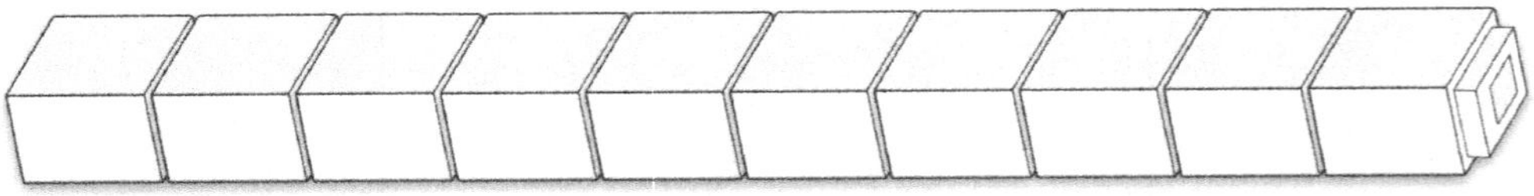

$10 - 6 = \square$ $\quad$ $10 - 1 = \square$ $\quad$ $10 - \square = 8$

$10 - 3 = \square$ $\quad$ $10 - 4 = \square$ $\quad$ $10 - \square = 5$

1.OA.4

Do you understand? **?** ✓

Affirmation: I am a mathematical thinker.

Name: ______________________________ Date: ______________

Unit 1 Practice 1: I can add and subtract within 10.

3) Molly has a cube tower. She breaks 6 cubes off of her cube tower and hides the other cubes behind her back. Write the subtraction equation she can use to model the number of cubes she has behind her back.

☐ - ☐ = ☐

1.OA.6

4) Draw a line to match the expression to its value.

5 + 2   7 - 2   7 - 5   4 + 2

5   7   6   2

1.OA.6

5) Find the sum, difference, or the missing number.

1 + 7 = ☐	9 - ☐ = 5
10 - 6 = ☐	7 - ☐ = 2
7 + 2 = ☐	3 + ☐ = 7

1.OA.8

Do you understand? ? ✓

Affirmation: I am brave and I take risks in math.

Name: ____________________ Date: ____________

Unit 1 Practice 2: I can relate addition and subtraction equations.

1) Use the counters to write an addition and a subtraction equation.

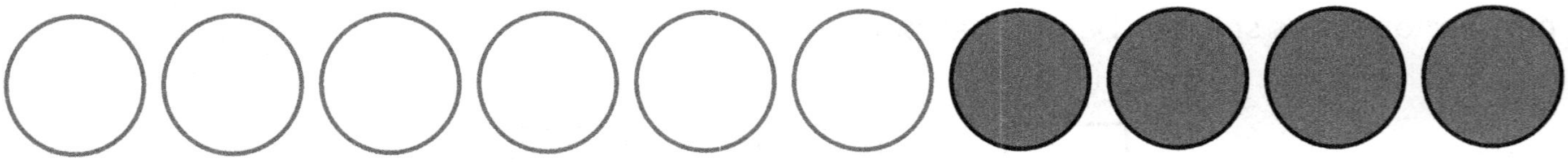

☐ + ☐ = ☐

☐ - ☐ = ☐

1.OA.4

2) Find the unknown. Match each subtraction equation to the addition equation with the same unknown.

9 - 3 = ☐	1 + ☐ = 6
10 - 7 = ☐	7 + ☐ = 10
6 - 1 = ☐	3 + ☐ = 9

1.OA.4

Do you understand? **?** ✓

Name: ______________________________ Date: ______________

Unit 1 Practice 2: I can relate addition and subtraction equations.

3) Color the cubes to model the addition equation.

$$10 = 2 + 8$$

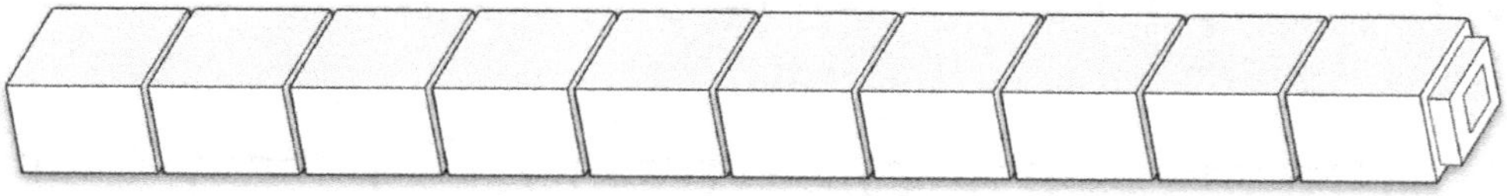

Write a related subtraction equation.

[ ] - [ ] = [ ]

2.OA.2

4) Use the number cards to create addition and subtraction equations that are true.

8 3 5

[ ] + [ ] = 8 [ ] - [ ] = 5

2.OA.2

Do you understand? ? ✓

Name: ______________________________ Date: ______________

Unit 1 Practice 3: I can find unknown numbers in equations within 20.

1) Match each expression to its value.

8 + 2	10 - 2	10 - 1	10 - 3	10 + 10	20 - 15
9	7	5	10	8	20

1.OA.6

2) Find the value that makes each equation true.

1 + ☐ = 20 16 + ☐ = 20

☐ - 19 = 1 ☐ - 16 = 4

15 + ☐ = 20 20-5 = ☐

1.OA.8

3) Write an addition and a subtraction equation that matches the image below.

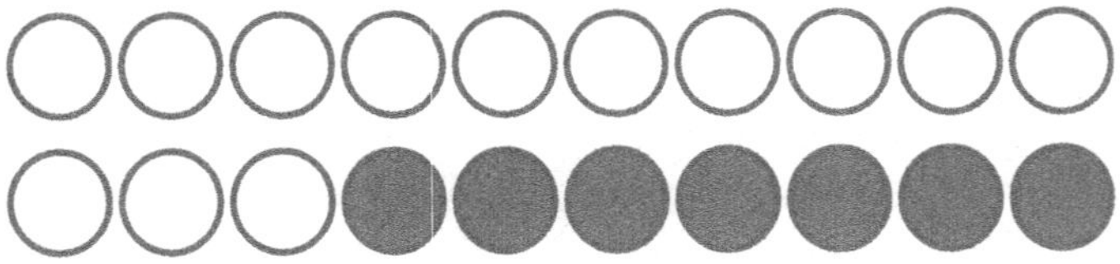

☐ + ☐ = ☐ ☐ - ☐ = ☐

1.OA.6

Do you understand? **?** ✓

Name: ______________________________ Date: ______________

Unit 1 Practice 3: I can find unknown numbers in equations within 20.

4) Jaquan has 20 cubes.
How many cube towers of 10 can he make?

Equation: ______________________________

2.OA.2

5) Sally has 18 pennies in a jar.
She takes 9 pennies out of the jar.
How many pennies are still in the jar?

Equation: ______________________________

2.OA.2

Do you understand? **?** ✓

Name: ______________________ Date: ____________

Unit 1 Practice 4: I can add and subtract within 20.

1) Find the **sum**:

$12 + 3 =$ ☐ $16 + 4 =$ ☐

$5 + 9 =$ ☐ $6 + 11 =$ ☐

$10 + 7 =$ ☐ $7 + 8 =$ ☐

$14 + 3 =$ ☐ $7 + 13 =$ ☐

1.OA.6

2) Find the **difference**:

$8 - 3 =$ ☐ $19 - 4 =$ ☐

$18 - 3 =$ ☐ $18 - 2 =$ ☐

$20 - 2 =$ ☐ $7 - 3 =$ ☐

$10 - 2 =$ ☐ $17 - 3 =$ ☐

1.OA.6

Do you understand? **?** ✓

Affirmation: When I do math, I feel like a superhero.

Name: ______________________ Date: ____________

Unit 1 Practice 4: I can add and subtract within 20.

3) Choose two cards below to make the addition and subtraction equations true. You may only use a card one time.

0 1 2 3 4

5 6 7 8 9

7 = ☐ + ☐

7 = ☐ + ☐

7 = ☐ - ☐

7 = ☐ - ☐

2.OA.2

4) Amy and Chris are playing, "How Close to 20?" Amy has the three number cards 6, 4, and 8. Chris has the three number cards 2, 9 and 7.

Amy ☐ + ☐ + ☐ = ☐

Chris ☐ + ☐ + ☐ = ☐

Who will win the game? ______________________

2.OA.1

Do you understand? **?** ✓

Name: ______________________________ Date: ______________

Unit 1 Practice 5: I can find sums to 50.

1) Count by 10s.

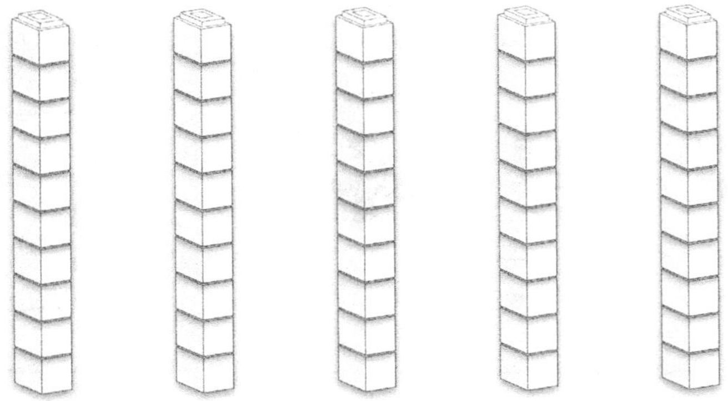

______, ______, ______, ______, ______

1.NBT.1

2) Draw cube towers and cubes to model the sum of 25 + 6.

Equation: ____

1.NBT.4

Do you understand? **?** ✓

Affirmation: It is important that I explain and justify my thinking.

Name: ______________________________ Date: ______________

Unit 1 Practice 5: I can find sums to 50.

3) Zoey has 3 cube towers and 6 cubes.
Her teacher gives her 1 cube tower and 3 cubes.
What number will be shown by the cube towers and cubes Zoey has now?

Equation: ____

1.NBT.4

4) Wyatt has cube towers and cubes shown below.
How many more cube towers and cubes will he need to reach 50?

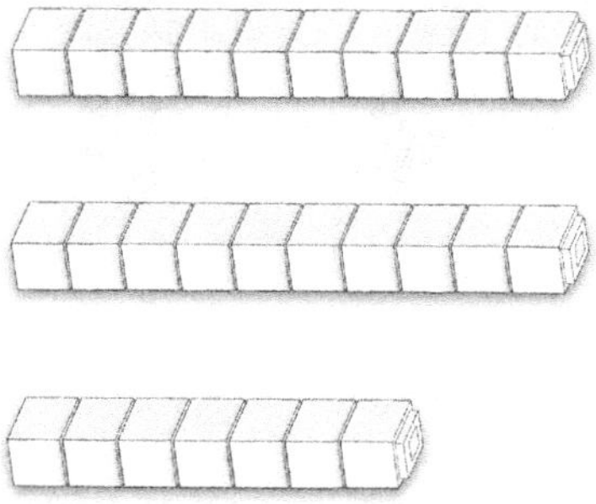

Wyatt needs _______ cube towers and _______ cubes to reach 50.

Equation: ____

2.NBT.5

Do you understand? **?** ✓

Name: ______________________ Date: ____________

Unit 1 Practice 6: I can practice addition and subtraction.

1) Eleven is one more than ten. Circle all of the expressions that have a value of 11.

12 - 2	10 + 1	20 - 8
20 - 9	4 + 7	8 + 3
2 + 9	6 + 5	13 - 3

1.OA.6

2) Circle true or false for each equation.

34 = 10 + 4

18 + 12 = 30

10 + 76 = 86

41 + 9 = 50

15 + 25 = 50

60 = 52 + 18

1.NBT.4

Do you understand? ? ✓

Name: ______________________________ Date: ______________

Unit 1 Practice 6: I can practice addition and subtraction.

3) Imani is playing 'Behind My Back' with 2 cube towers.
She breaks off 3 cubes and keeps the rest behind her back.
How many cubes does Imani have behind her back?

Equation: ______________________________

2.OA.2

4) Find the sum of 12 + 24 + 8.

Equation: ______________________________

2.NBT.6

Do you understand? **?** ✓

Affirmation: Math helps me make sense of my world.

Name: ______________________________ Date: ______________

Unit 1 Practice 7: I can organize and share data.

1) Write an equation to find the sum of the dots.

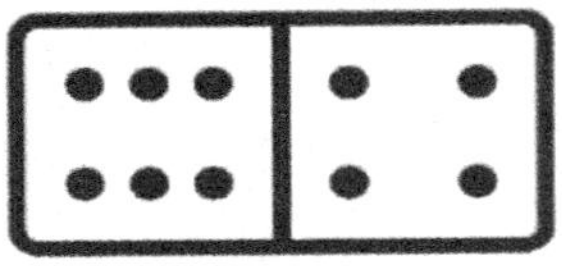

_______ + _______ = _______

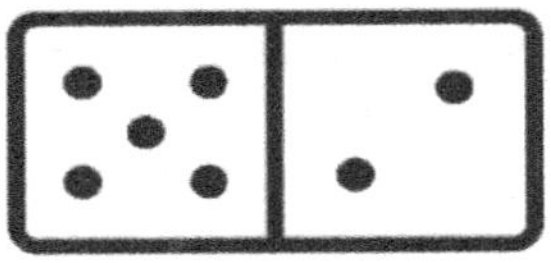

_______ + _______ = _______

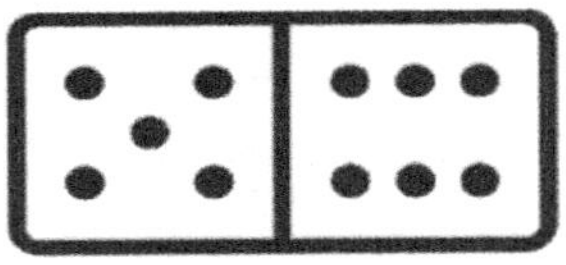

_______ + _______ = _______

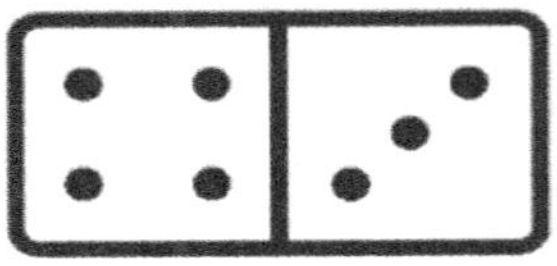

_______ + _______ = _______

1.OA.5

2) Ethan made a table to show how many of each shape he has.

How many total shapes does Ethan have?

Shape	Number of Shapes
Triangles	7
Circles	10
Squares	3

Ethan has _______ shapes.

Equation: ____

1.OA.2

Do you understand? **?** ✓

Name: ______________________________ Date: ______________

Unit 1 Practice 7: I can organize and share data.

Questions A and B refer to the information below.

Animals at the Zoo

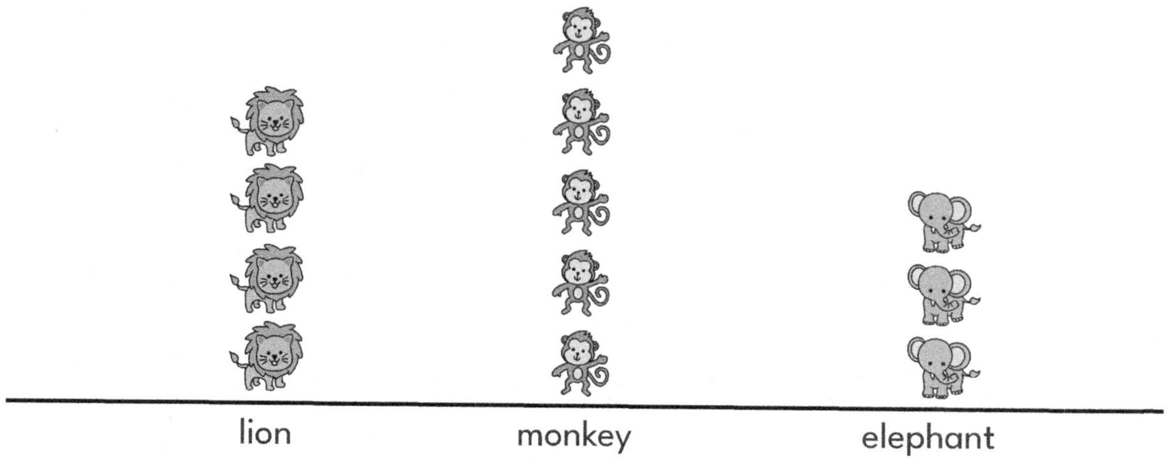

A) What kinds of animals are at this zoo?

There are ______________, ______________, and ______________ at the zoo.

1.MD.4

B) How many total animals are at this zoo?

There are ______ animals at this zoo.

Equation: ____

1.MD.4

Do you understand? **?** ✓

Name: ______________________________ Date: ______________

Unit 1 Practice 8: I can answer questions based on a picture graph.

1) Find the sum.

_______ + _______ + _______ = _______

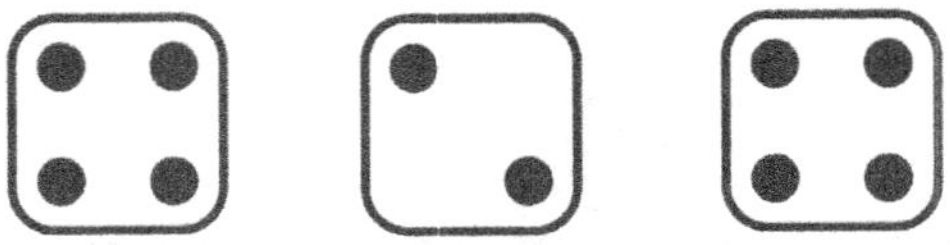

_______ + _______ + _______ = _______

1.OA.2

2) Madison made a chart showing the art supplies she counted in her class.

How many total art supplies did Madison count?

Madison counted _______ total art supplies.

Art Supplies	Number of Each
Pencils	15
Markers	10
Crayons	20

Equation: ____

1.OA.2

Do you understand? **?** ✓

Name: ______________________________ Date: ______________

Unit 1 Practice 8: I can answer questions based on a picture graph.

Questions A and B refer to the information below.

Animals at the Zoo

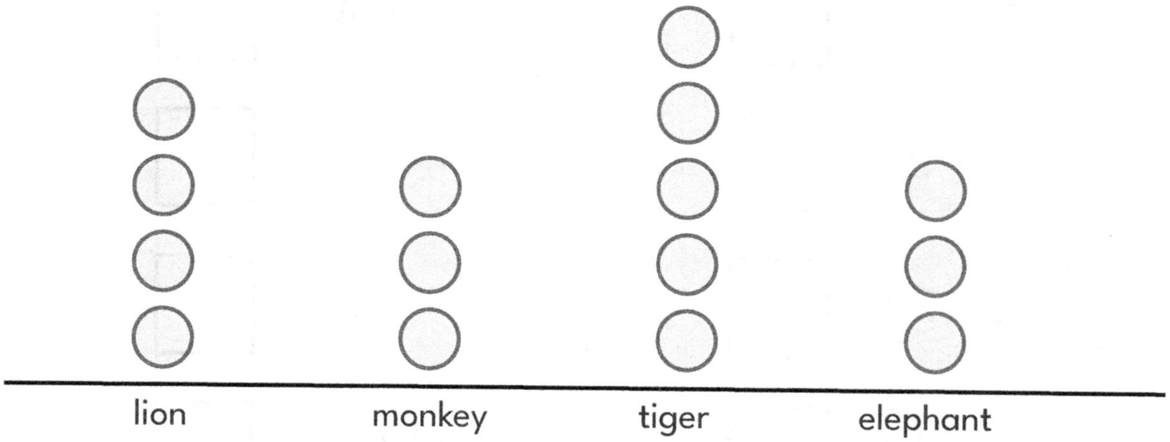

A) How many of each zoo animal does the picture graph show?

There are _______ lions.

There are _______ tigers.

There are _______ monkeys.

There are _______ elephants.

2.MD.10

B) How many lions and tigers are there at the zoo?

There are _______ lions and tigers at the zoo.

Equation: ____

2.MD.10

Do you understand? **?** ✓

Affirmation: Sketching out my thinking helps me see a problem more clearly.

Name: ______________________ Date: ____________

Unit 1 Practice 9: I can use a bar graph to answer questions about data.

1) Find the sum or difference:

4 + 16 = ☐	21 + 4 = ☐
15 + 3 = ☐	11 + 5 = ☐
10 - 7 = ☐	9 + 8 = ☐
14 - 3 = ☐	13 + 8 = ☐

2) Match the addition expressions to the subtraction expressions with the same value.

2 + 13	20 - 8
10 + 4	17 - 3
6 + 6	19 - 2
8 + 9	20 - 5

1.OA.5

Do you understand? **?** ✓

Name: ____________________ Date: __________

Unit 1 Practice 9: I can use a bar graph to answer questions about data.

Questions A and B refer to the graph below.

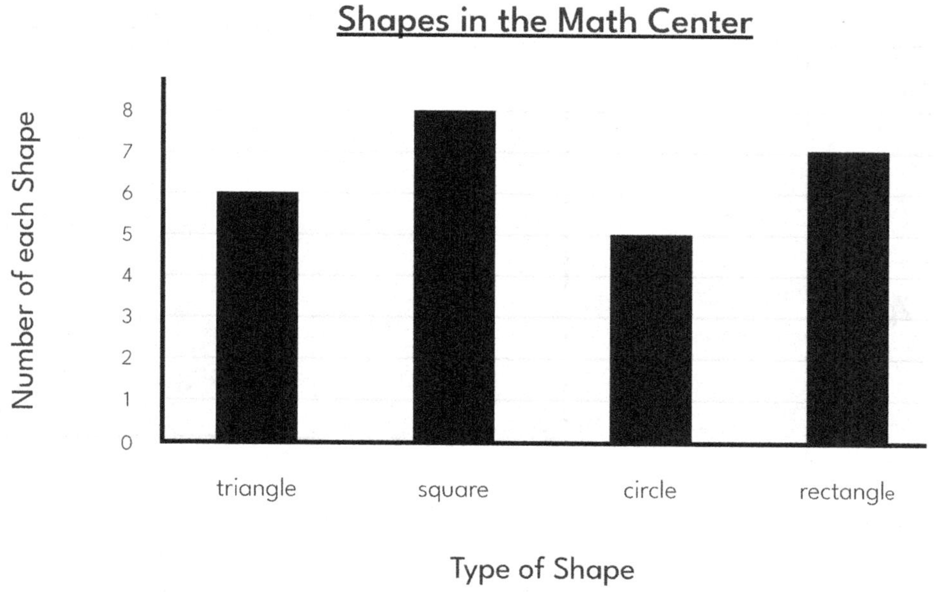

A) How many squares and rectangles are there in the math center?

There are _______ squares and rectangles in the math center.

Equation: ____________________

2.MD.10

B) How many **total** shapes are there in the math center?

There are _______ total shapes in the math center.

Equation: ____________________

2.MD.10

Do you understand? **?** ✓

Affirmation: Making mistakes is how I learn new things.

Name: ______________________________ Date: ______________

Unit 1 Practice 10: I can make a picture graph and a bar graph.

1) Use the table to make the bar graph.

Shape	Number of Each
triangle	6
square	4
circle	8
rectangle	3

Shapes in the Math Center

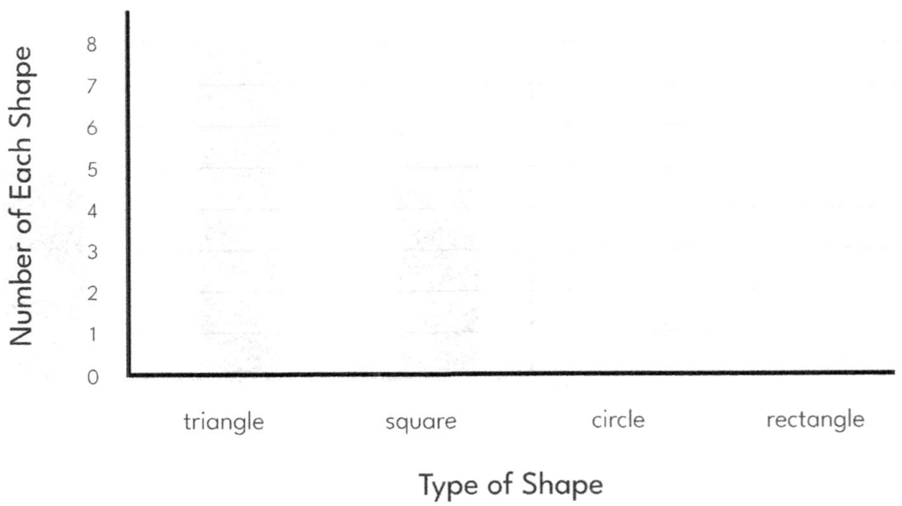

2) How many circles and triangles are there in the math center?

There are _______ circles and triangles in the math center.

Equation: ____

2.MD.10

3) How many total shapes are there in the math center?

There are _______ total shapes in the math center.

Equation: ____

2.MD.10

Do you understand? **?** ✓

Name: ____________________ Date: __________

Unit 1 Practice 10: I can make a picture graph and a bar graph.

Questions A and B refer to the information below.

Use the data from A to fill in the pictograph.

Animals at the Zoo

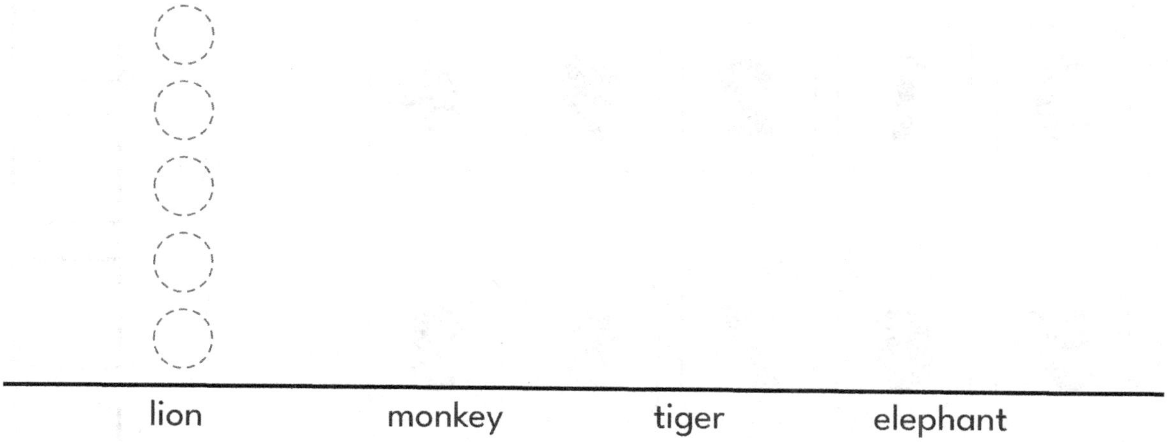

A) Sort the zoo animals.

There are __5__ lions.

There are ______ tigers.

There are ______ monkeys.

There are ______ elephants.

2.MD.10

B) Display the data in the picture graph above.

How many total animals are at the zoo?

There are ______ total animals at the zoo.

Equation: ____

2.MD.10

Do you understand? **?** ✓

Name: ______________________________ Date: ______________

Unit 1 Practice 11: I can answer questions about data.

1) Choose two cards below to make the addition and subtraction equations true.
You may only use a card one time.

0	1	2	3	4
5	6	7	8	9

8 = ☐ + ☐

8 = ☐ + ☐

8 = ☐ - ☐

8 = ☐ - ☐

2.OA.2

2) Circle true (thumbs up) or false (thumbs down) for each equation.

$10 = 5 + 4 + 5$

$10 + 10 + 10 = 30$

$21 + 2 + 6 = 29$

$15 + 15 + 10 = 40$

1.OA.2

Do you understand? ? ✓

Name: ______________________ Date: ____________

Unit 1 Practice 11: I can answer questions about data.

Questions A and B refer to the graph below.

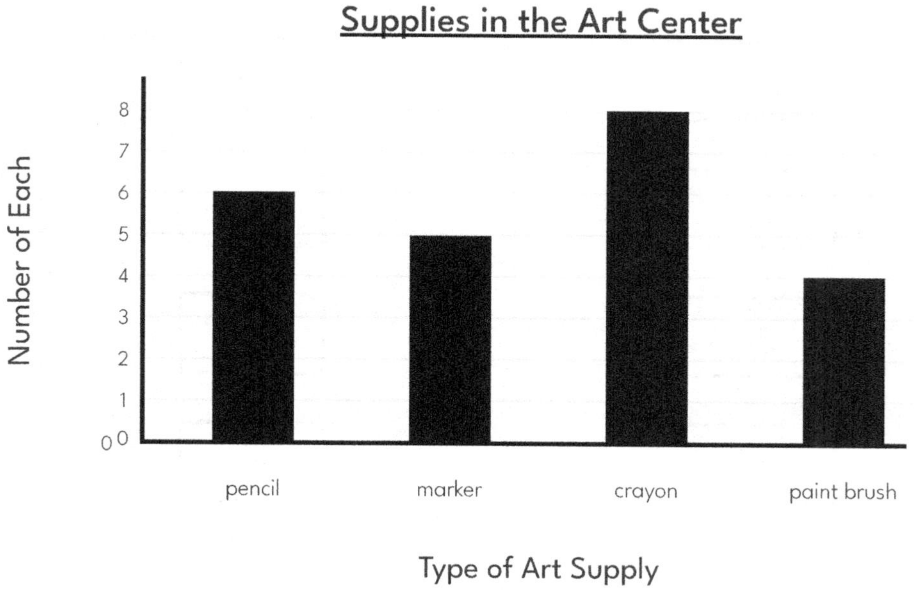

A) How many pencils, markers and crayons are in the art center?

There are _______ pencils, markers and crayons in the art center.

Equation: ____

2.MD.10

B) Which supply is there the most of?

The art center has the most ______________________.

Which supply is there the fewest of?

The art center has the fewest ______________________.

2.MD.10

Do you understand? ? ✓

Name: ______________________________ Date: ______________

Unit 1 Practice 12: I can make a graph to display data.

1) Match each expression to its difference.

20 - 4	13
19 - 6	16
24 - 5	12
20 - 8	19

2.OA.2

2) Write an addition and a subtraction equation that matches the image below.

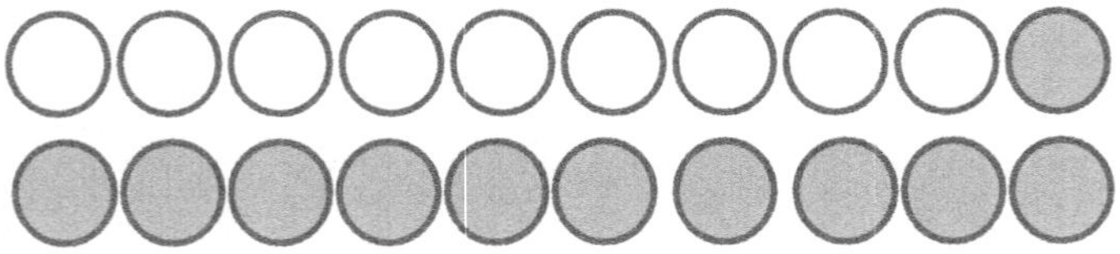

☐ + ☐ = ☐   ☐ - ☐ = ☐

1.OA.5

Do you understand? **?** ✓

Name: ______________________________ Date: ______________

Unit 1 Practice 12: I can make a graph to display data.

3) Use the table to make the bar graph.

Zoo Animal	Number of Student Votes
lion	8
monkey	6
tiger	4
elephant	7

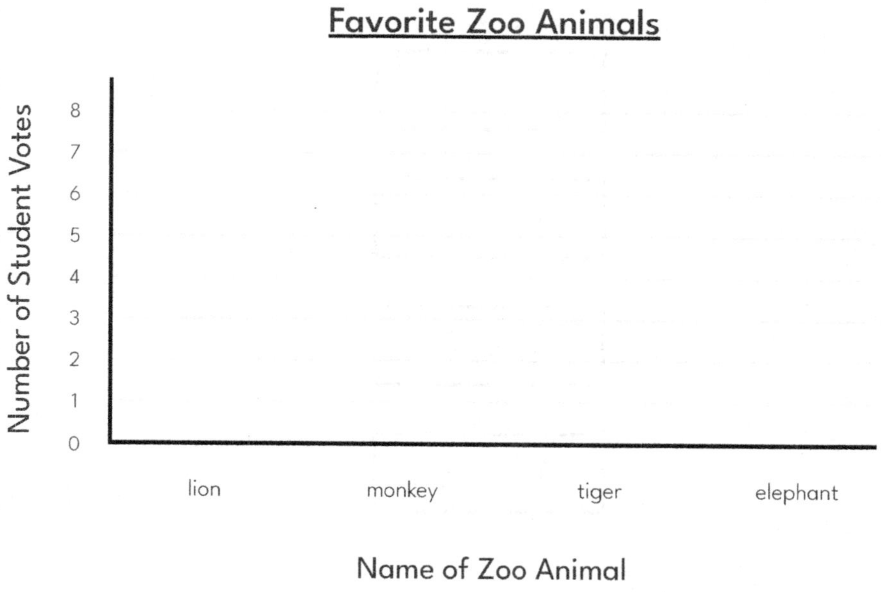

4) How many students voted for tigers? _______ students voted for tigers.

How many students voted for elephants? _______ students voted for elephants.

2.MD.10

5) How many more students voted for elephants than for tigers?

_______ more students voted for elephants than for tigers.

Equation: ____

2.MD.10

Do you understand? **?** ✔

Name: ______________________________ Date: ______________

Unit 1 Practice 13: I can solve compare problems using a bar graph.

1) Match the expression with the **same value**.

9 + 3	11 + 10
14 + 9	7 + 10
8 + 9	13 + 10
12 + 9	10 + 2

2) Find the sum of three **addends**.

7 + 8 + 3 = ☐	10 + 5 + 10 = ☐
6 + 10 + 4 = ☐	15 + 5 + 5 = ☐
1 + 4 + 9 = ☐	6 + 6 + 10 = ☐
2 + 6 + 2 = ☐	10 + 9 + 9 = ☐

1.OA.2

Do you understand? **?** ✓

Name: ______________________________ Date: ______________

Unit 1 Practice 13: I can solve compare problems using a bar graph.

Questions A and B refer to the graph below.

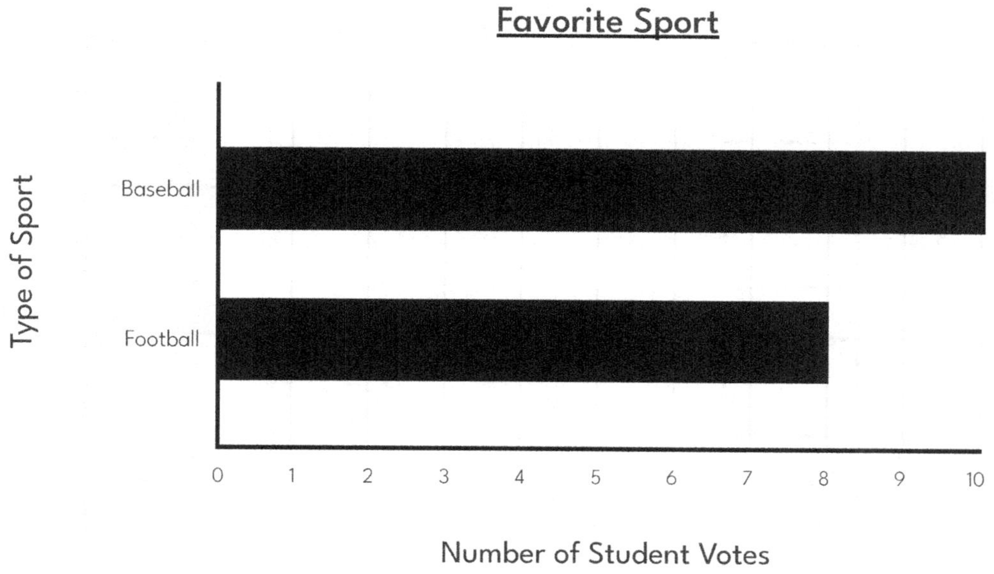

A) Which sport got the most student votes?

__________________ got the most student votes.

Which sport got the fewest student votes?

__________________ got the fewest student votes.

2.MD.10

B) How many more students voted for baseball than football?

_______ more students voted for baseball than for football.

Equation: ____

2.MD.10

Do you understand? **?** ✓

Name: ____________________ Date: __________

Unit 1 Practice 14: I can solve compare problems using a tape diagram.

1) Find the difference.
Subtract a one-digit number from a two-digit number.

17 - 5 = ☐	19 - 4 = ☐
27 - 5 = ☐	49 - 4 = ☐
16 - 3 = ☐	12 - 3 = ☐
36 - 3 = ☐	22 - 3 = ☐

1.NBT.4

2) Find the difference.
Subtract a multiple of ten from a two-digit number.

10 - 10 = ☐	35 - 10 = ☐
13 - 10 = ☐	35 - 20 = ☐
23 - 10 = ☐	56 - 20 = ☐
33 - 10 = ☐	46 - 20 = ☐

1.NBT.4

Do you understand? ? ✓

Affirmation: If I messed up yesterday, today is a new day, and I can try again.

Name: ______________________________ Date: ______________

Unit 1 Practice 14: I can solve compare problems using a tape diagram.

Questions A and B refer to the information below.

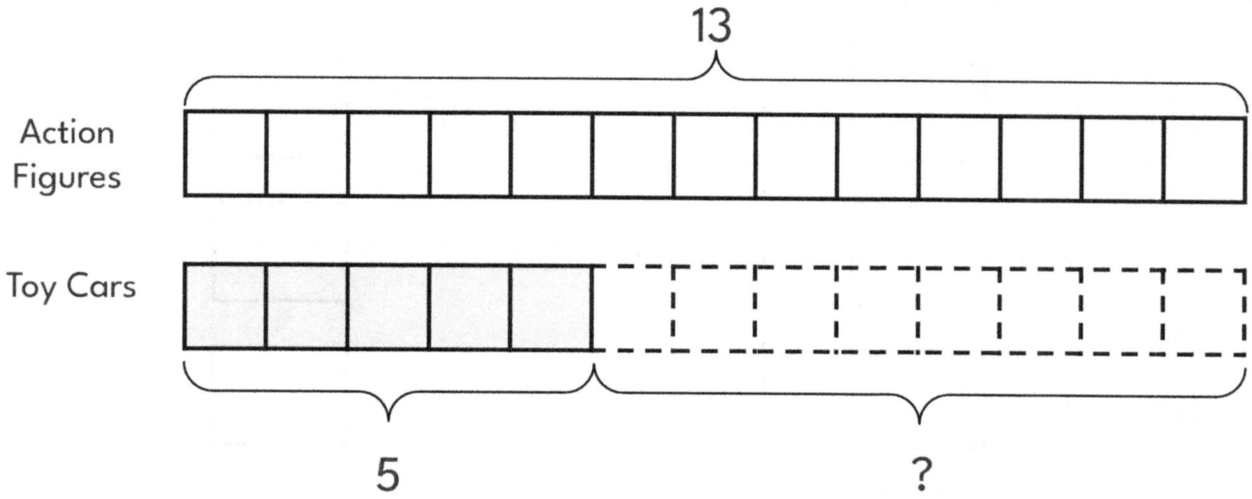

A) The diagram shows how many action figures and toy cars Liam has. How many action figures does Liam have?

Liam has _______ action figures.

How many toy cars does Liam have?

Liam has _______ toy cars.

2.MD.10

B) How many more action figures than toy cars does Liam have?

Liam has _______ more action figures than toy cars.

Equation: ____

2.MD.10

Do you understand? **?** ✓

Name: ______________________________ Date: ______________

Unit 1 Practice 15: I can solve compare problems with unknowns in all positions.

1) Find the difference.
   Subtract a two-digit number from a two-digit number.

18 - 12 = ☐	25 - 13 = ☐
28 - 12 = ☐	52 - 11 = ☐
38 - 12 = ☐	39 - 19 = ☐
48 - 12 = ☐	20 - 19 = ☐

1.NBT.4

2) Hanna has 20 triangles and 12 circles.
   How many more triangles than circles does Hanna have?

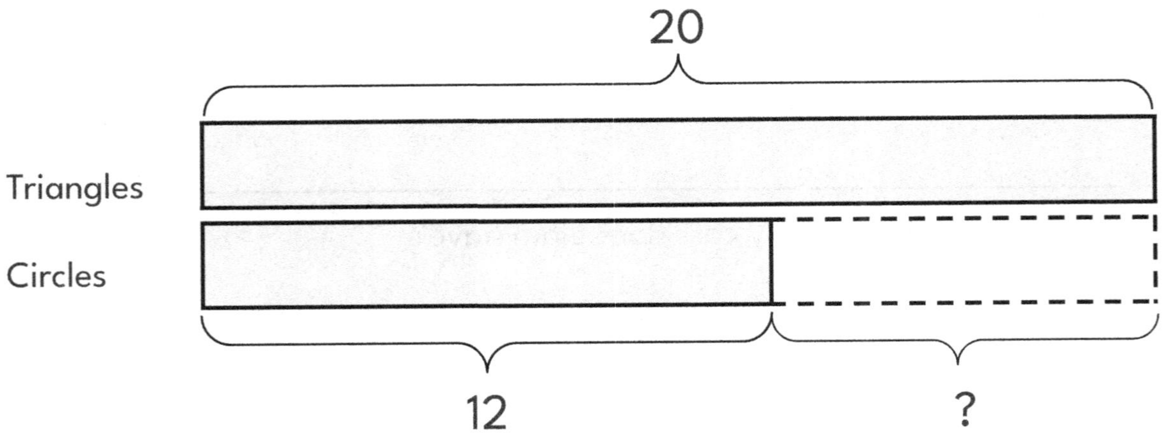

Use the tape diagram to write an equation and solve the problem.

Hanna has _______ more triangles. Equation: ______________________________

2.OA.1

Do you understand? ? ✓

Name: ______________________________ Date: ____________

Unit 1 Practice 15: I can solve compare problems with unknowns in all positions.

3) Logan has 25 baseball cards. Ivy has 12 fewer baseball cards than Logan.
How many baseball cards does Ivy have?

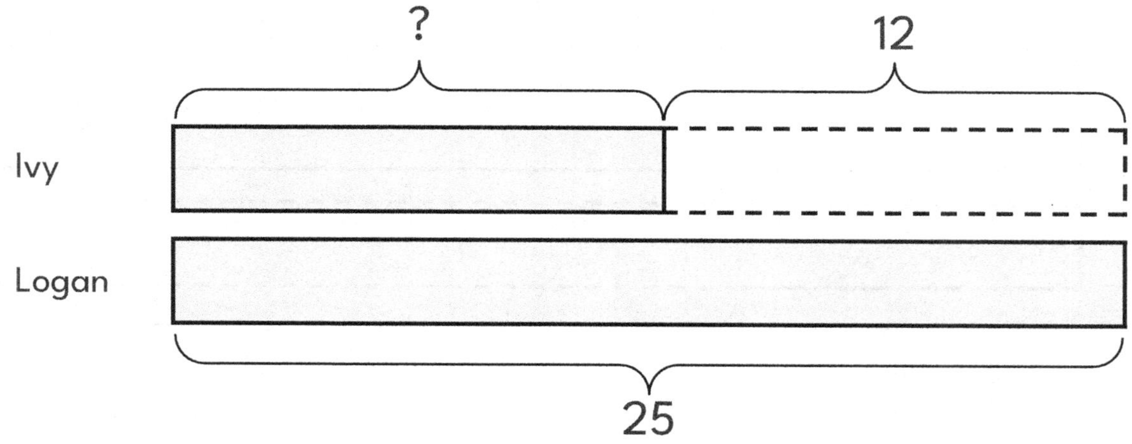

Use the tape diagram to write an equation and solve the problem.

Ivy has _______ baseball cards. Equation: ______________________________

2.OA.1

4) Ethan has 46 marbles in his collection. Chloe has 13 fewer marbles than Ethan.
How many marbles does Chloe have in her collection?
Complete the tape diagram.

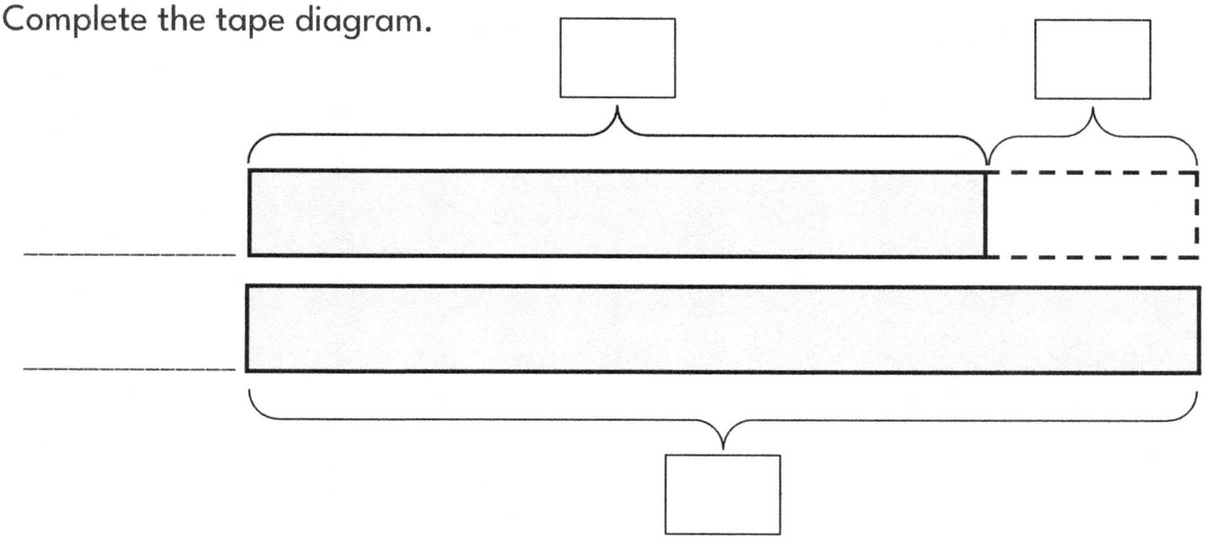

Chloe has _______ marbles in her collection. Equation: ______________________________

2.OA.1

Do you understand? ? ✓

Name: ______________________________ Date: ______________

Unit 1 Practice 16: I can solve compare problems using a tape diagram.

1) There are 32 red tiles and 12 green tiles.
Complete the tape diagram

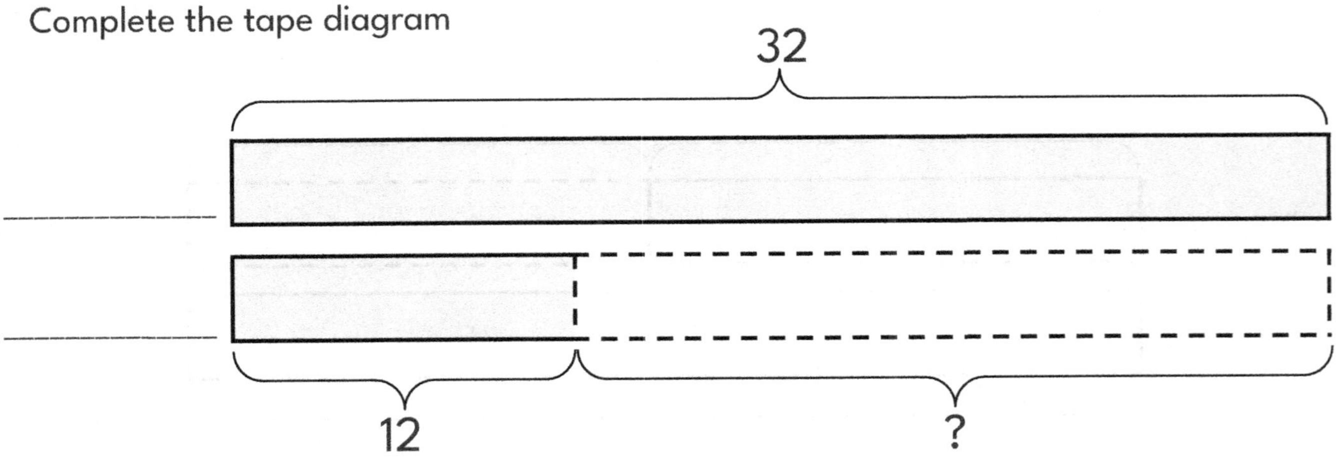

How many more red tiles are there than green tiles?

There are _______ more red tiles. Equation: ______________________________

2.OA.1

2) There are 46 triangles and 22 circles in the shape center.
How many fewer circles are there than triangles?
Draw a tape diagram.

There are _______ fewer circles. Equation: ______________________________

2.OA.1

Do you understand? **?** ✓

Name: ______________________________ Date: ______________

Unit 1 Practice 16: I can solve compare problems using a tape diagram.

Questions A and B refer to the graph below.

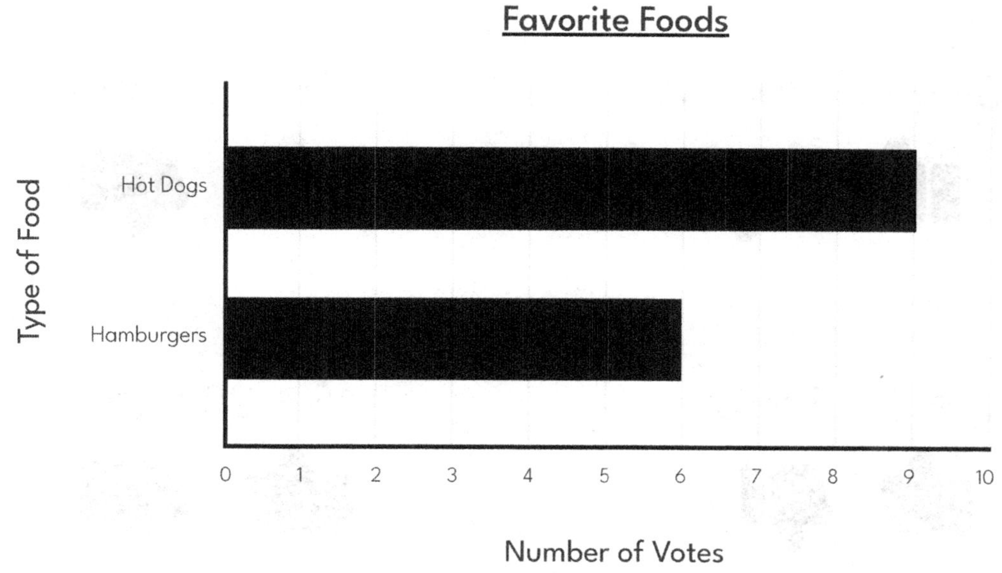

A) How many more people voted for hot dogs than hamburgers? (Do not solve)
Create a tape diagram to model the problem.

2.MD.10

B) How many more people voted for hot dogs than hamburgers?
Write an equation and solve.

Equation: ______________________________

2.MD.10

Do you understand? **?** ✓

Affirmation: Think positive thoughts. Negative thoughts don't help us learn and grow.

Name: ______________________________ Date: ______________

Unit 1 Practice 17: I can estimate and count collections.

1) Circle true or false for each equation.

$34 + 10 + 10 = 54$

$28 = 18 + 10$

$42 + 5 + 5 = 52$

$18 + 20 = 28$

2.NBT.2

2) Sam wants to represent 57 base-ten blocks.
He has created this collection of base-ten blocks.
Sam know that he needs more.

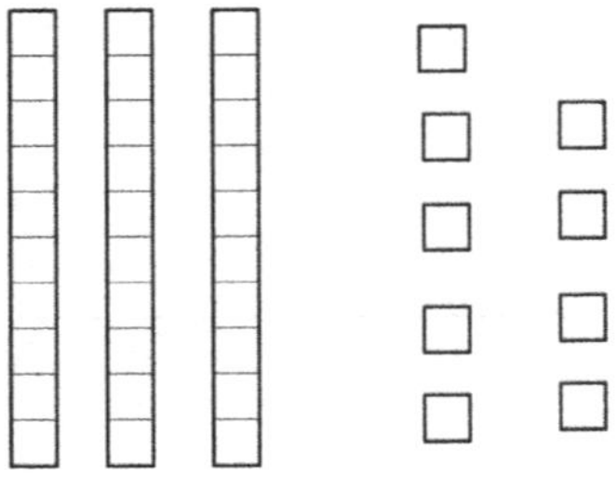

How many more base-ten blocks does Sam need to complete his collection?

Sam needs _______ more base-ten blocks.

Draw base-ten blocks to complete Sam's representation.

2.NBT.5

Do you understand? ? ✓

Name: ______________________________ Date: ______________

Unit 1 Practice 17: I can estimate and count collections.

Questions A and B refer to the information below.

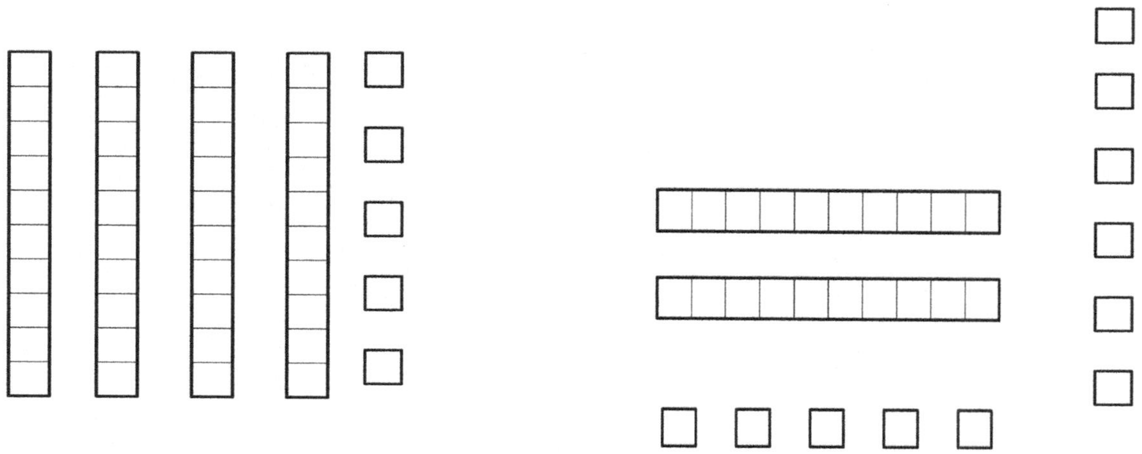

A) Elana estimated that there are 90 base-ten blocks in the collection.
Is her estimate too low, too high or about right?

Elana's estimate is ______________________________.

2.NBT.2

B) Count the collection of base-ten blocks.

There are _______ base-ten blocks in the collection.

How did you count the collection?

2.NBT.2

Do you understand? **?** ✓

Name: ______________________________ Date: ______________

Unit 1 Practice 18: I can complete a picture graph and a bar graph using data.

Questions A and B refer to the graph below.

Title: ______________________________

Number of Pets

__________ __________ __________ __________

Different Pets

A) Complete the picture graph using this data that tells about pets in a pet store.

There are 5 birds.

There are 3 dogs.

There are 7 lizards.

There are 2 cats.

2.MD.10

B) How many more lizards are there than birds in the pet store?

There are _______ more lizards than birds in the pet store.

Equation: ______________________________

2.MD.10

Do you understand? ? ✓

Name: ____ Date: _______________

Unit 1 Practice 18: I can complete a picture graph and a bar graph using data.

Questions A and B refer to the graph below.

<u>Tiles in the Math Center</u>

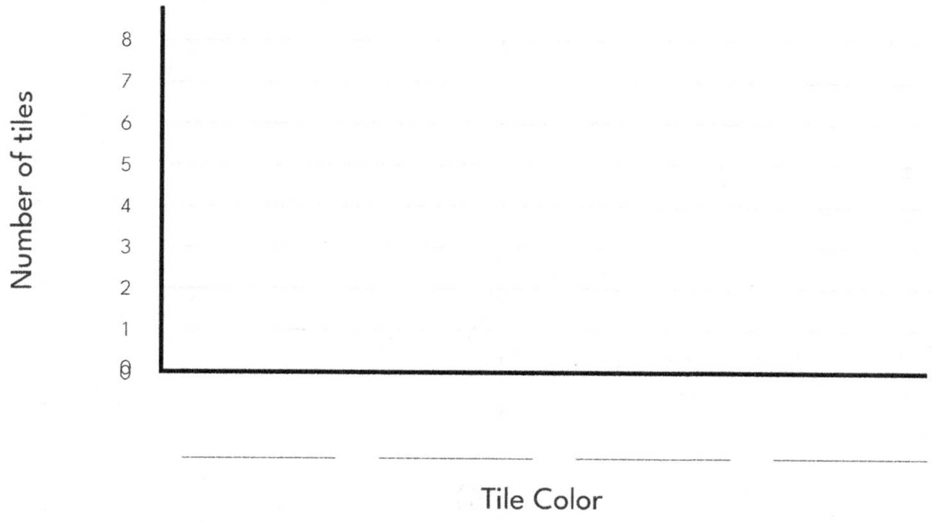

A) Complete the bar graph using this data about different color tiles in the math center.

There are 8 red tiles.

There are 2 blue tiles.

There are 7 green tiles.

There are 4 yellow tiles.

2.MD.10

B) How many tiles are there in the math center?

There are _______ tiles in the math center.

Equation: ____

2.MD.10

Do you understand? **?** ✓

Name: ______________________________ Date: ______________

I can practice grade level fluencies.

Set 1: Find the **sum.**	**Set 2:** Find the **difference.**
1) 8 + 5 = _______	1) 20 - 6 = _______
2) 9 + 6 = _______	2) 20 - 4 = _______
3) 12 + 8 = _______	3) 15 - 5 = _______
4) 4 + 15 = _______	4) 15 - 6 = _______
5) 3 + 17 = _______	5) 24 - 3 = _______
6) 17 + 4 = _______	6) 21 - 2 = _______
7) 9 + 8 = _______	7) 17 - 6 = _______
2.NBT.5	2.NBT.5

Do you understand? **?** ✓

Unit Reflection

Adding, Subtracting, and Working with Data

Use this space to reflect on your understanding of the unit skills and concepts.

Skill/Concept	I can...	I need to work on...

Unit 2

Adding and Subtracting within 100

In this unit, we will learn ways to add and subtract numbers up to 100 by thinking about place value and how numbers work together. We will use these strategies to solve different kinds of story problems.

Place Value with Tens and Ones
The number 58 has 5 tens 8 ones

Addition and Subtraction
$12 + 7 = 19$
$19 - 7 = 12$

Use Base Ten Blocks to Decompose a Ten

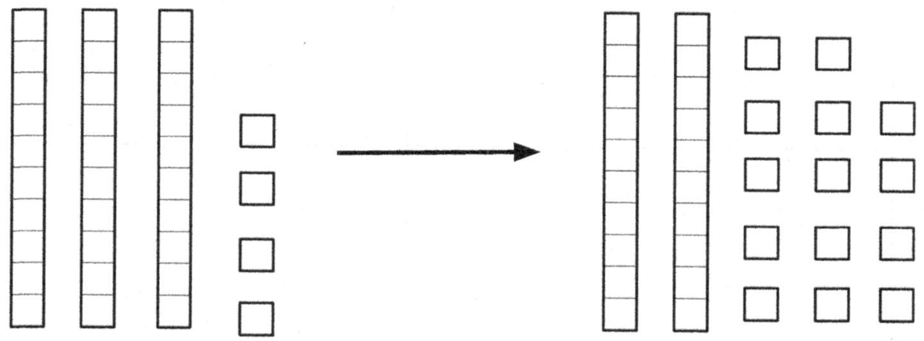

3 tens 4 ones **2 tens 14 ones**

Colin has 47 action figures. Eric has 74 action figures. How many more action figures does Eric have than Colin?

Unit Vocabulary

Adding and Subtracting within 100

Use this space to visualize the math vocabulary for this unit.

Word or Phrase	Example or Attributes	Visual Reminder

Unit Models & Strategies

Adding and Subtracting within 100

Use this space to visualize the math models and strategies for this unit.

Model or Strategy	This is a...	It is used to...

Affirmation: I am human and humans are mathematical beings.

Name: ________________________________ Date: ______________

Unit 2 Practice 1: I can add and subtract using different models.

1) Mary and Jake built collections using cube trains and cubes.

Mary

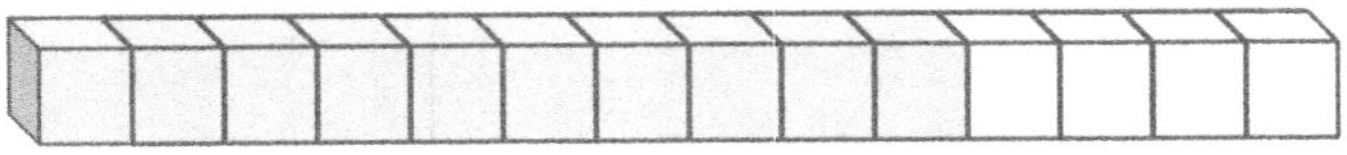

Jake

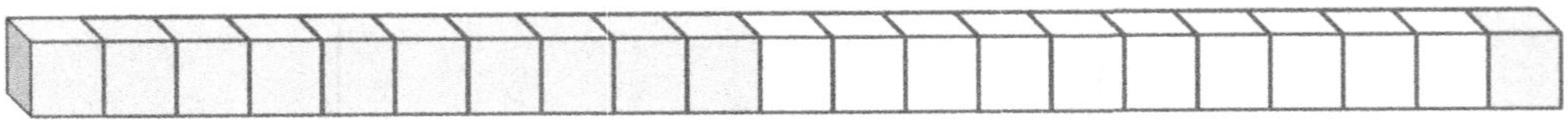

How many cubes in Mary's collection?

There are _______ cubes in Mary's collection.

How many cubes in Jakes collection?

There are _______ cubes in Jake's collection.

2.NBT.2

2) Paul used base-ten models to represent his sports card collection.
How many of each does Paul have in his collection?

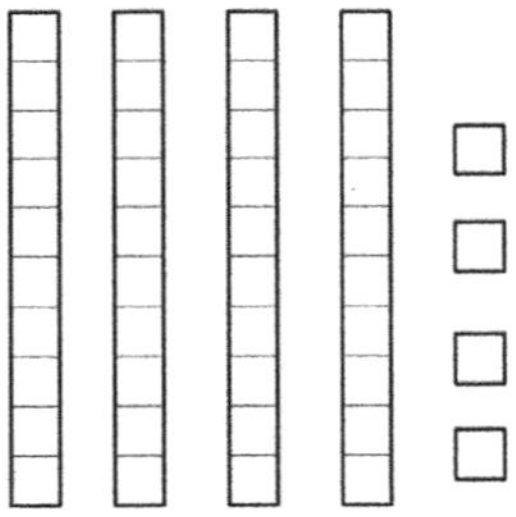

baseball cards

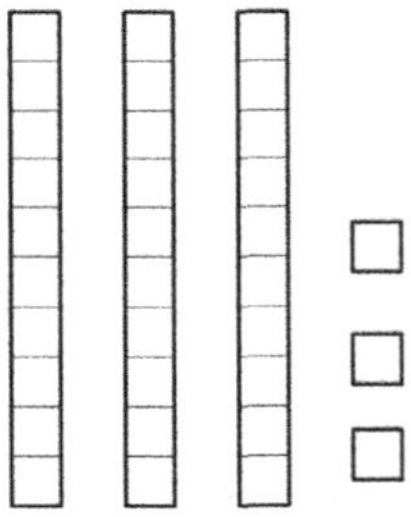

football cards

There are _______ baseball cards in Paul's collection.

There are _______ football cards in Paul's collection.

2.NBT.2

Do you understand? **?** ✓

Name: ______________________________ Date: ______________

Unit 2 Practice 1: I can add and subtract using different models.

Questions A and B refer to the graph below.

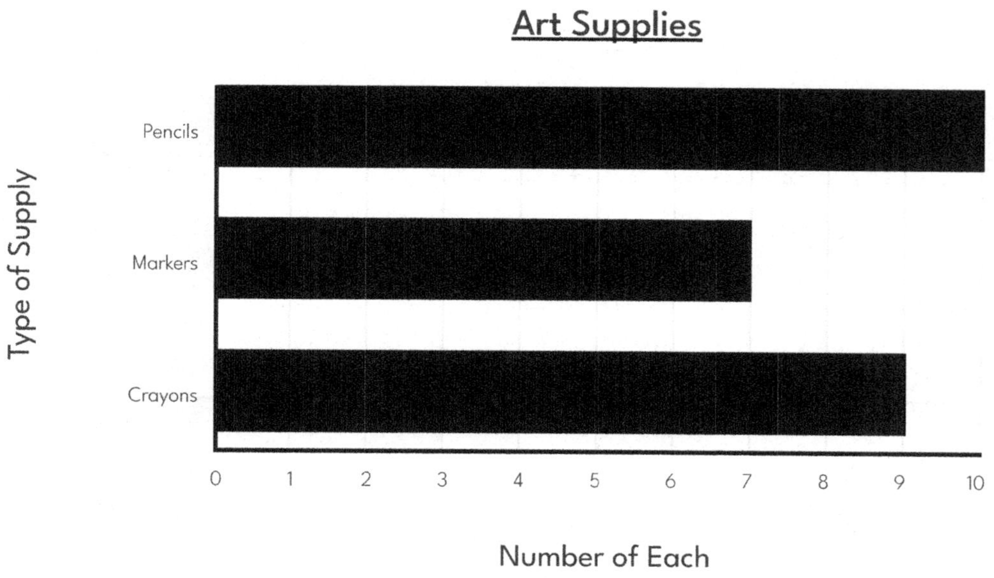

A) How many markers and crayons are there in the art center?

There are _______ markers and crayons in the art center.

Equation: ______________________________

2.MD.10

B) How many more pencils than crayons are in the art center?

There are _______ more pencils.

Equation: ______________________________

2.MD.10

Do you understand? **?** ✓

Name: ______________________________ Date: ______________

Unit 2 Practice 2: I can use addition or subtraction to find an unknown addend.

1) Count on by tens.

14, ______, ______, ______, ______, ______, ______, ______, 94

35, ______, ______, ______, ______, ______, ______, ______, 115

Count back by tens.

90, ______, ______, ______, ______, ______, ______, ______, 10

93, ______, ______, ______, ______, ______, ______, ______, 13

1.NBT.5

2) Match the number that makes each equation true.

28 + ☐ = 48	10
43 + ☐ = 53	40
67 - ☐ = 37	20
82 - ☐ = 42	30

1.OA.8

Do you understand? **?** ✓

Affirmation: Even when we know the answer, it can be fun to try out a different way.

Name: ______________________________ Date: ______________

Unit 2 Practice 2: I can use addition or subtraction to find an unknown addend.

3) Justin has this collection of red cubes.

red 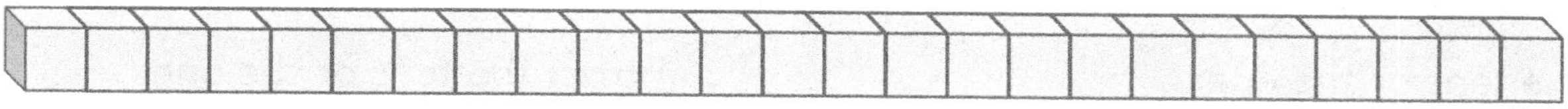

blue

Justin wants to add some blue cubes so that his total collection will have 47 cubes.

Draw the blue cubes he will need.

Justin needs _______ blue cubes.

Equation: ______________________________

2.NBT.5

4) Solve: 56 + _______ = 96

Equation: ______________________________

2.NBT.5

Do you understand? **?** ✓

Name: ______________________________ Date: ______________

Unit 2 Practice 3: I can solve story problems using addition or subtraction.

1) Find the sum or difference.

Add the tens or the ones.

56 + 10 = _______

56 + 20 = _______

56 + 30 = _______

47 + 1 = _______

47 + 2 = _______

Subtract the tens or the ones.

56 - 10 = _______

56 - 20 = _______

56 - 30 = _______

56 - 1 = _______

56 - 2 = _______

1.NBT.4

2) Isaac had a box of 72 crayons.
He gave some to his brother.
Isaac now has 52 crayons.
How many crayons did Isaac give to his brother?

Equation: ____

2.OA.1

Do you understand? **?** ✓

Name: ______________________________ Date: ______________

Unit 2 Practice 3: I can solve story problems using addition or subtraction.

3) Sally is cleaning the art center.
She collected all of the markers from the art center.
She put 36 of the markers in a box and 30 markers in a bag.
How many markers did Sally collect from the art center?

Equation: ______________________________

2.OA.1

4) Nora has 47 crayons and 24 markers.
How many more crayons than markers does Nora have?

Equation: ______________________________

2.OA.1

Do you understand? **?** ✓

Name: ______________________________ Date: ______________

Unit 2 Practice 4: I can add or subtract using tens and ones.

1) **Sample:** Subtract tens then subtract ones to solve two-digit subtraction.

Solve: 37 - 12

Base-ten Model:

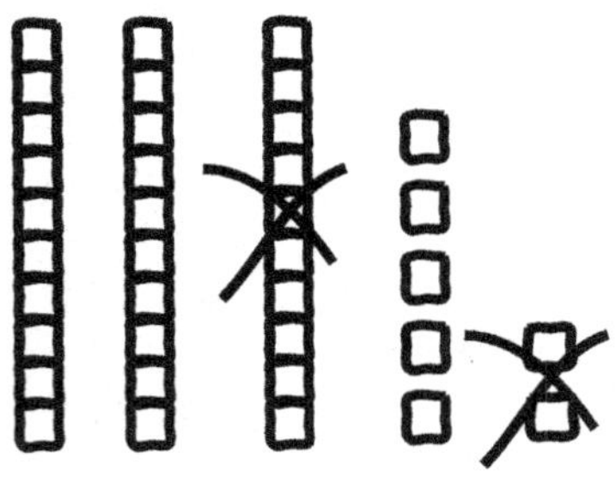

Subtract the tens: 30 - 10 = 20

Subtract the ones: 7 - 2 = 5

Add to solve: 20 + 5 = 25

Equation: 37 - 12 = 25

2.NBT.5

2) Find the difference.

Solve: 46 - 31

Base-ten Model:

Subtract the tens: ______________________

Subtract the ones: ______________________

Add to solve: ______________________

Equation: ____

2.NBT.5

Do you understand? **?** ✓

Name: ______________________________ Date: ______________

Unit 2 Practice 4: I can add or subtract using tens and ones.

3) Find the sum: 43 + 25 = ______

Solve by adding the tens then adding the ones.

Equation: ____

2.NBT.5

4) Find the difference: 68 - 23

Solve by subtracting the tens then subtracting the ones.

Equation: ____

2.NBT.5

Do you understand? **?** ✓

Name: ____________________ Date: __________

Unit 2 Practice 5: I can subtract a one-digit number from a two-digit number.

1) Count back by ones.

32, ____, ____, 29, ____, ____, ____, ____, ____, 23

45, ____, ____, ____, ____, ____, 39, ____, ____, 36

27, ____, ____, ____, ____, ____, ____, ____, 19, ____, 17

64, ____, ____, ____, ____, 59, ____, ____, ____, 55

1 OA.5

2) Counting back to subtract.
Use the count back by ones work you just completed to find the difference.

32 - 2 = ____ 27 - 7 = ____

32 - 3 = ____ 27 - 8 = ____

45 - 5 = ____ 64 - 4 = ____

45 - 6 = ____ 64 - 5 = ____

1.OA.5

Do you understand? ? ✓

Name: ______________________________ Date: ______________

Unit 2 Practice 5: I can subtract a one-digit number from a two-digit number.

3) Count back to solve.
Use the cube towers and cubes to find the difference.

17 - 9

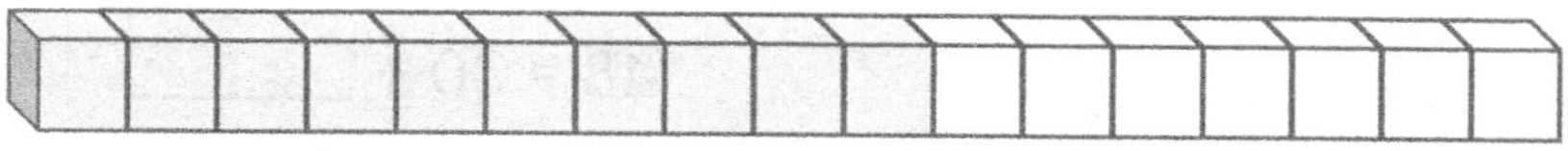

Equation: ______________________________

2.OA.2

4) Count back to solve.
Use the cube towers and cubes to find the difference.

24 - 7

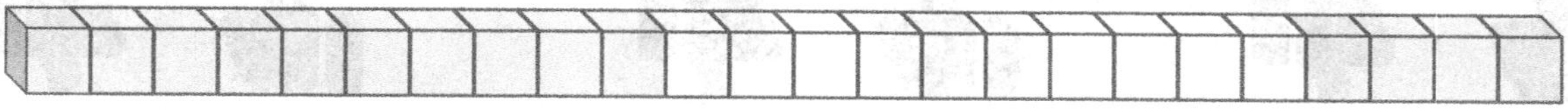

Equation: ______________________________

2.OA.2

Do you understand? **?** ✓

Affirmation: New solutions happen when we try new ways of doing things.

Name: ______________________ Date: ____________

Unit 2 Practice 6: I can decompose a ten to subtract.

1) Write the number that makes each equation true.

31 = 30 + ______ 48 = 40 + ______

31 = 20 + ______ 48 = 30 + ______

57 = 50 + ______ 25 = 20 + ______

57 = 40 + ______ 25 = 10 + ______

2.OA.2

2) True or False : You can subtract without breaking apart a ten.

23 - 2 45 - 5 45 - 8

57 - 6 98 - 9 32-5

2.OA.2

Do you understand? **?** ✓

Affirmation: New solutions happen when we try new ways of doing things.

Name: ______________________________ Date: ______________

Unit 2 Practice 6: I can decompose a ten to subtract.

3) Saul is solving 49 - 8 using a base-ten diagram.
Show how Saul can use the base-ten diagram to find the difference.

49 - 8 = ________

Did Saul need to decompose a ten?

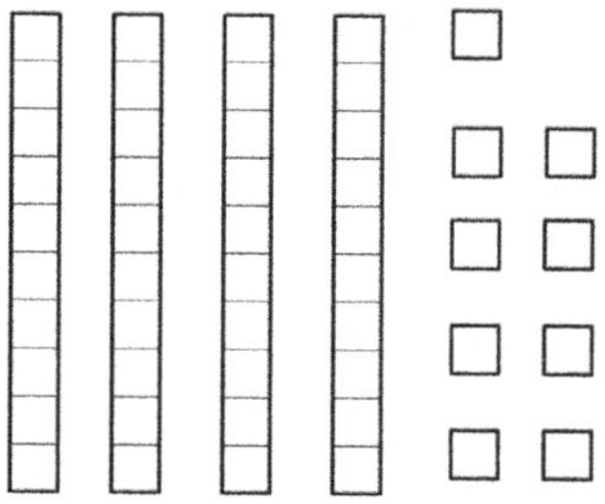

2.NBT.5

4) Polly is solving 58 - 9 using a base-ten diagram.
Show how Polly can use the base-ten diagram to find the difference.

58 - 9 = ________

Did Polly need to decompose a ten?

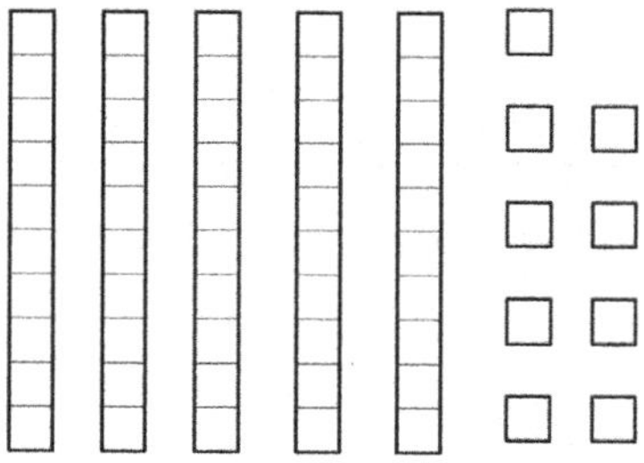

2.NBT.5

Do you understand? **?** ✓

Name: ______________________________ Date: ______________

Unit 2 Practice 7: I can compose or decompose a ten.

1) **Sample:** Use the base-ten model to compose another ten.

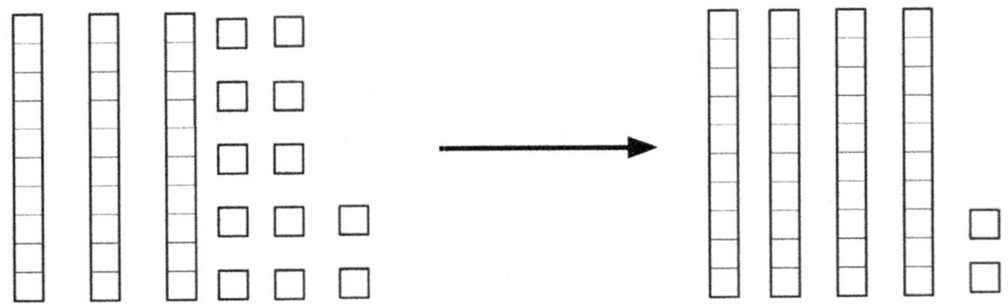

3 tens 12 ones → 4 tens 2 ones

1.NBT.2

2) Use the base-ten model to compose another ten.

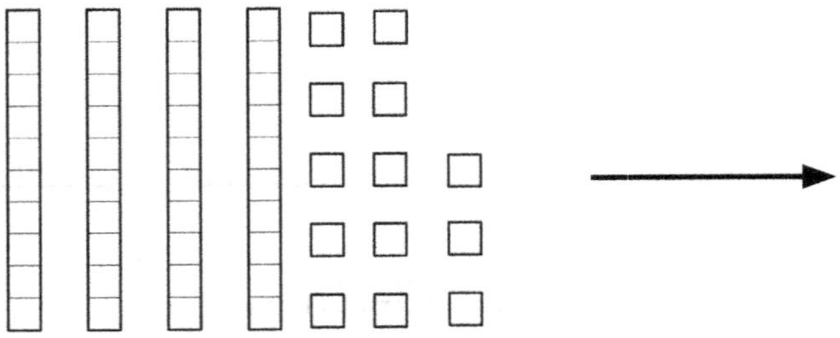

_______ tens _______ ones → _______ tens _______ ones

1.NBT.2

3) Use the base-ten model to compose another ten.

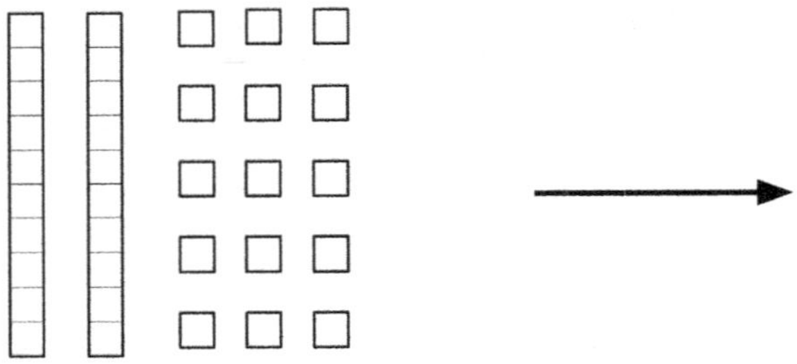

_______ tens _______ ones → _______ tens _______ ones

1.NBT.2

Do you understand? **?** ✓

Affirmation: By asking questions, I learn more and help others learn as well.

Name: ______________________________ Date: ______________

Unit 2 Practice 7: I can compose or decompose a ten.

4) **Sample:** Use the base-ten model to decompose a ten.

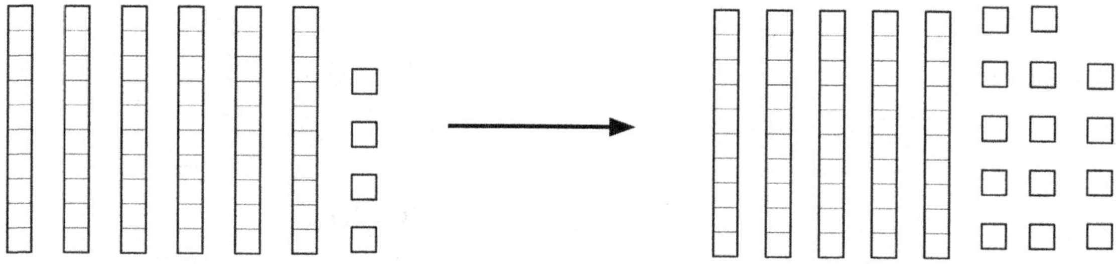

6 tens 4 ones → 5 tens 14 ones

1.NBT.2

5) Use the base-ten model to decompose a ten.

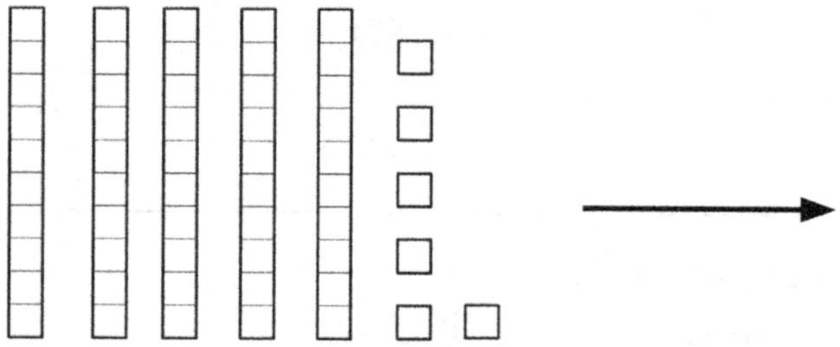

________ tens ________ ones ________ tens ________ ones

1.NBT.2

6) Use the base-ten model to decompose a ten.

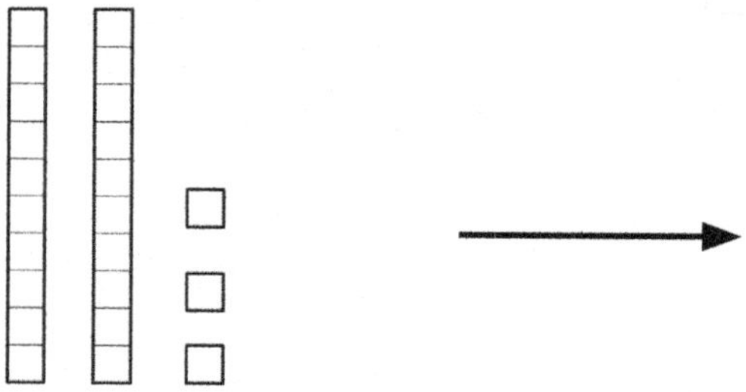

________ tens ________ ones ________ tens ________ ones

1.NBT.2

Do you understand? ? ✓

Affirmation: I love learning! I am a learner.

Name: ______________________________ Date: ______________

Unit 2 Practice 8: I can use models or equations to subtract.

1) Write the number that makes each equation true.

62 = 50 + ☐   23 = 20 + ☐

62 = 40 + ☐   23 = 10 + ☐

37 = 30 + ☐   91 = 90 + ☐

37 = 20 + ☐   91 = 80 + ☐

2.OA.2

2) Frank is using using a base-ten diagram to subtract 32 - 9.
Show how Frank can use the base-ten diagram to find the difference.

32 - 9 = ________

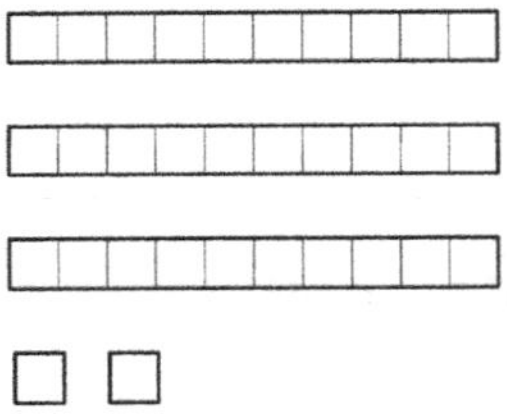

Did Frank need to decompose a ten?

2.NBT.5

Do you understand? **?** ✓

Name: ______________________________ Date: ______________

Unit 2 Practice 8: I can use models or equations to subtract.

3) Hazel used equations to subtract 53 - 28.
Hazel wrote:

$$53 - 20 = 33$$

$$33 = 20 + 13$$

$$13 - 8 = 5$$

$$20 + 5 = 25$$

Explain how Hazel used equations to subtract.

2.NBT.9

4) Violet also used equations to subtract 53 - 28.
Violet wrote:

$$53 = 40 + 13$$

$$40 - 20 = 20$$

$$13 - 8 = 5$$

$$20 + 5 = 25$$

Explain how Violet used equations to subtract.

2.NBT.9

Do you understand? ? ✓

Affirmation: Everyone can become better at math! Including ME!

Name: ______________________________ Date: ______________

Unit 2 Practice 9: I can add or subtract using two-digit numbers

1) Circle true or false for each equation.

97 = 81 + 10 + 7

82 = 60 + 10 + 10 + 2

46 = 40 + 16

32 = 10 + 10 + 10 + 2

1.NBT.2

2) Match the number that makes each equation true.

32 - 12 =	50
32 - 13 =	20
63 - 13 =	49
63 - 14 =	19

2.NBT.5

Do you understand? **?** ✓

Name: ______________________________ Date: ______________

Unit 2 Practice 9: I can add or subtract using two-digit numbers.

3) Find the sum: 38 + 18

Equation: ______________________________

2.NBT.5

4) Find the difference: 42 - 13

Equation: ______________________________

2.NBT.5

Do you understand? **?** ✓

Name: ______________________________ Date: ______________

Unit 2 Practice 10: I can find differences with two-digit subtraction.

1) Subtract multiples of ten.
Write the number that makes each equation true.

42 - 10 = ☐	78 - 10 = ☐
42 - 20 = ☐	78 - 20 = ☐
42 - 30 = ☐	78 - 30 = ☐
42 - 40 = ☐	78 - 40 = ☐

1.NBT.6

2) Victoria is playing the game, "*Too Low, Too High, About Right.*"
Her guess is that this collection shows 6 tens and 4 ones.
Is Victoria Too Low, Too High, or About Right?

Victoria's guess is:

❑ Too Low

❑ Too High

❑ About Right

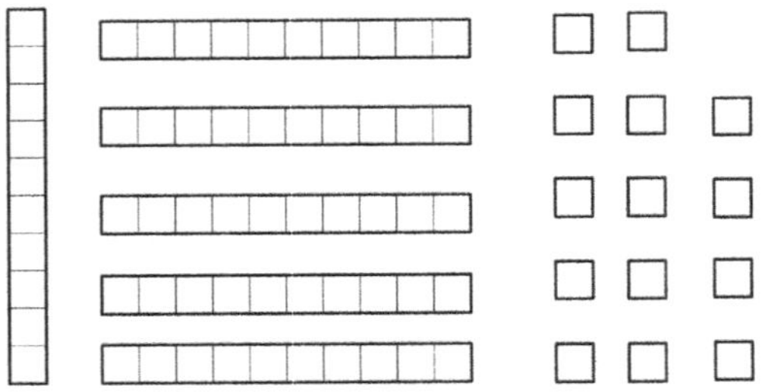

1.NBT.2

Do you understand? **?** ✓

Affirmation: I can do harder math problems by drawing out the story.

Name: ______________________ Date: ____________

Unit 2 Practice 10: I can find differences with two-digit subtraction.

3) Solve 64 - 27 using a base-ten diagram.

Equation: ______________________

2.NBT.5

4) Solve 53 - 48 using equations.

Equation: ______________________

2.NBT.5

Do you understand? ? ✓

Name: ________________________ Date: ____________

Unit 2 Practice 11: I can solve story problems using two-digit numbers.

1) Ezekiel has a collection of 67 sports cards.
He has 32 baseball cards and the rest are football cards.
How many football cards does Ezekiel have in his collection?

Ezekiel has _______ football cards in his collection.

1.OA.1

2) Ezekiel had a collection of 67 sports cards.
He got a new packet of 12 sports cards.
How many sports cards does Ezekiel have now?

Ezekiel now has _______ sports cards in his collection.

1.OA.1

Do you understand? **?** ✓

Name: ______________________________ Date: ______________

Unit 2 Practice 11: I can solve story problems using two-digit numbers.

3) Sophia had a collection of 60 baseball cards.
She gave 28 of the cards to her sister.
How many baseball cards does Sophia have now?

Sophia now has _______ baseball cards in her collection.

2.NBT.5

4) Maya has a collection of 74 football cards.
Thomas has 25 fewer football than Maya in his collection.
How many football cards does Thomas have?

Thomas has _______ football cards in his collection.

2.OA.1

Do you understand? **?** ✓

Affirmation: Everyone, I mean everyone makes mistakes, especially in math.

Name: ______________________ Date: ____________

Unit 2 Practice 12: I can use diagrams to solve story problems.

1) Caleb has a sticker collection.
He has 19 dinosaur stickers and 43 action figure stickers.
How many stickers does Caleb have in his collection?
Use the tape diagram to solve.

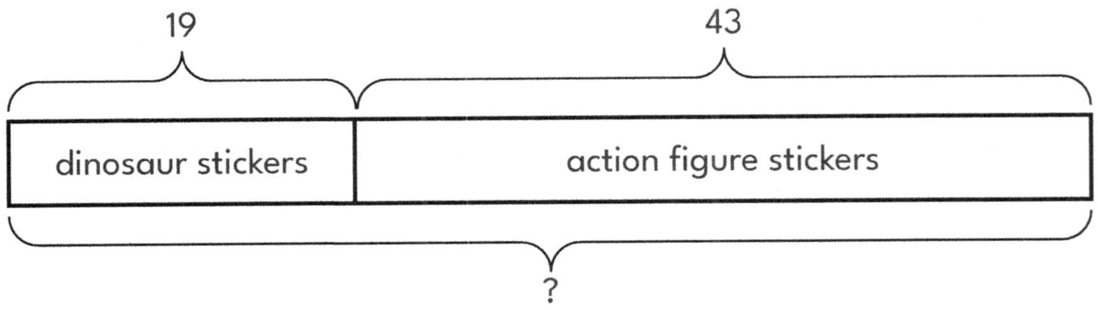

Caleb has _______ stickers in his collection.

2.OA.1

2) Halona has a collection of 47 animal stamps.
She has 25 dog stamps and the rest are turtle stamps.
How many turtle stamps does Halona have in her collection?
Use the tape diagram to solve.

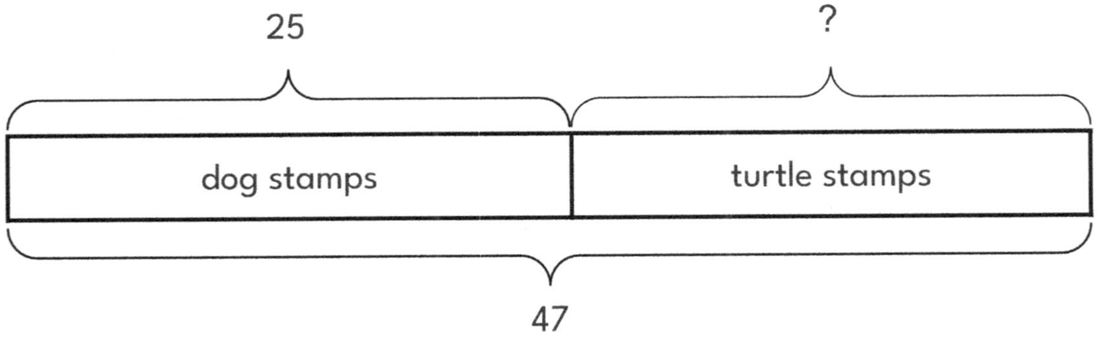

Halona has _______ turtle stamps in her collection.

2.OA.1

Do you understand? ? ✓

Name: ______________________________ Date: ______________

Unit 2 Practice 12: I can use diagrams to solve story problems.

3) Violet has a star sticker collection.
She has 27 red star stickers and 12 blue star stickers.
How many stickers does Violet have in her collection?
Draw a tape diagram to solve.

Violet has _______ stickers in her collection.

2.OA.1

4) Camila has a collection of 60 zoo animal stamps.
She has 30 zebra stamps and the rest are tiger stamps.
How many tiger stamps does Camila have in her collection?
Draw a tape diagram to solve.

Camila has _______ tiger stickers in her collection.

2.OA.1

Do you understand? **?** ✓

Affirmation: I am smart in my own way and so is everyone else.

Name: ______________________ Date: ____________

Unit 2 Practice 13: I can solve story problems and write equations.

1) Circle the equations that match the tape diagram.

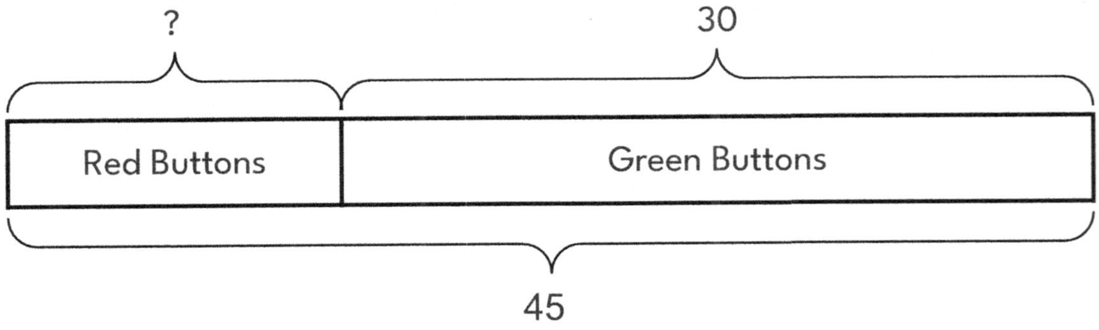

30 + 45 = ?	? + 30 = 45	45 = ? + 30

2.OA.1

2) Emma has 39 baseball cards and some soccer trading cards.
She has a total of 60 trading cards.
Complete the tape diagram to match the equation.

39 + ? = 60

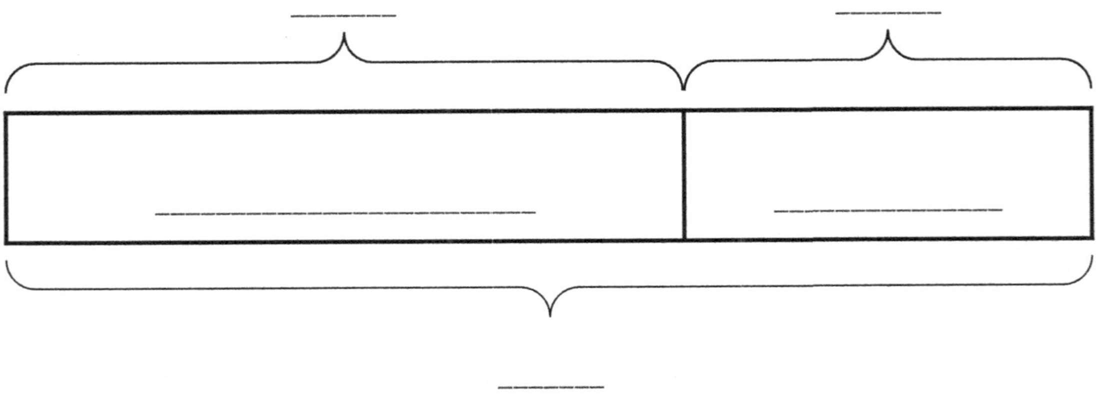

2.OA.1

Do you understand? **?** ✓

Affirmation: I am smart in my own way and so is everyone else.

Name: ______________________________ Date: ______________

Unit 2 Practice 13: I can solve story problems and write equations.

3) Theo has 28 stickers. Carl has 15 fewer stickers than Theo.
Complete the compare tape diagram.

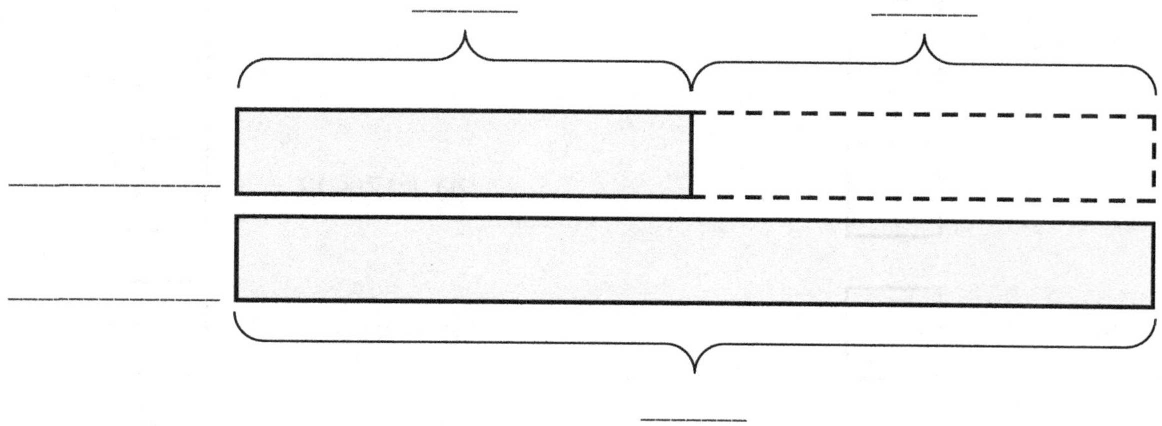

Equation with a question mark for the unknown: ______________________

2.OA.1

4) Alice has a collection of action figure stickers. She has 80 blue action figure stickers. She has 50 fewer red action figure stickers than blue action figure stickers.
Complete the tape diagram.

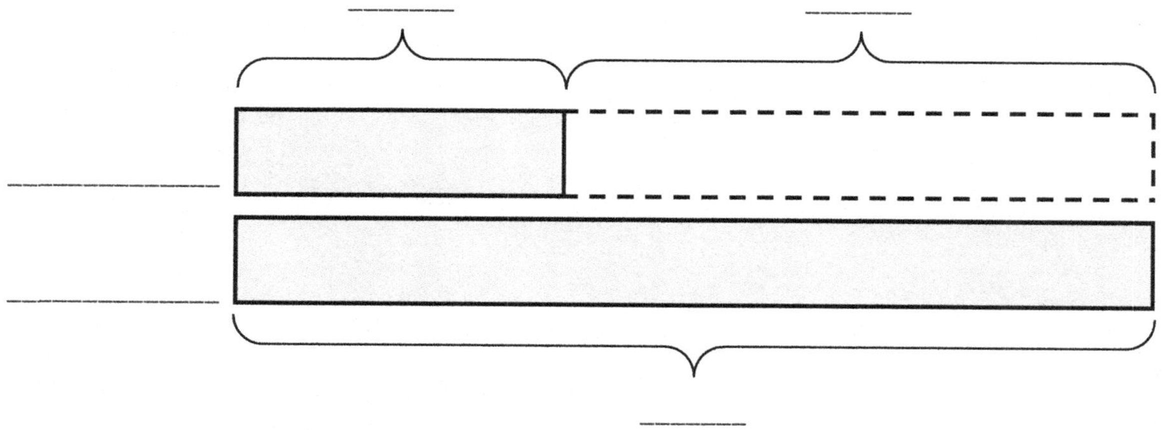

Equation with a question mark for the unknown: ______________________

2.OA.1

Do you understand? **?** ✓

Affirmation: I can use anything as a math manipulative.

Name: ______________________________ Date: ____________

Unit 2 Practice 14: I can solve story problems with three addends.

1) Find sums with three addends.
   Write the number that makes each equation true.

6 + 7 + 4 = ☐	3 + 10 + 7 = ☐
10 + 20 + 30 = ☐	20 + 17 + 13 = ☐
12 + 12 + 12 = ☐	38 + 30 + 12 = ☐
5 + 10 + 5 = ☐	25 + 25 + 25 = ☐

1.OA.2

2) Matthew's zoo animal stamp collection has 22 zebra stamps, 32 tiger stamps and 16 lion stamps. How many zoo animal stamps does Matthew have in his collection?

Matthew has _______ zoo animal stamps in his collection.

2.OA.1

Do you understand? **?** ✓

Name: ______________________________ Date: ______________

Unit 2 Practice 14: I can solve story problems with three addends.

3) Noah and Caleb have sports card collections. Noah's collection has 30 baseball cards and 17 football cards. Caleb's collection has 25 baseball cards and zero football cards. How many sports cards do the two boys have together?

Noah and Caleb have _______ sports cards in their combined collections.

2.OA.1

4) Noah is adding 35 soccer sports cards to his collection. He already has 30 baseball cards and 17 football cards. How many sports cards will Noah have in his expanded collection?

Noah now has _______ sports cards in his expanded collection.

2.OA.1

Do you understand? **?** ✓

Name: ____________________ Date: ____________

Unit 2 Practice 15: I can use a diagram to write a story problem.

1) Match the expressions that have the same difference.

28 - 12	40 - 25
35 -20	29 - 13
50 - 25	88 - 44
94 - 50	75 - 50

2.NBT.5

2) Peter is playing *"Target Number."* He started at 100 and the target number is zero. For his first turn, Peter subtracted 30. For his second turn, Peter subtracted 40. What will his starting number be for his third turn?

For his third turn, Peter's starting number will be _______ .

1.NBT.6

Do you understand? **?** ✓

Name: ______________________________ Date: ______________

Unit 2 Practice 15: I can use a diagram to write a story problem.

Questions A and B refer to the information below.

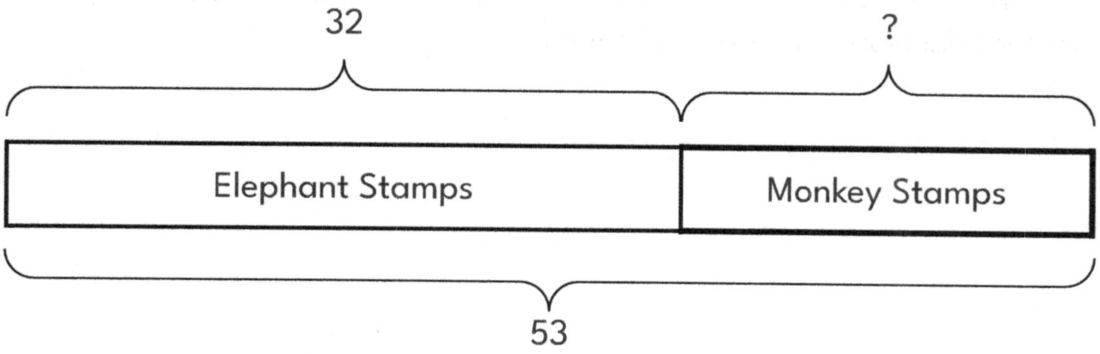

A) Use the tape diagram to write a story problem.

2.OA.1

B) Solve your story problem.

Write an equation with a question mark for the unknown.

Equation: ______________________________

2.OA.1

Do you understand? **?** ✓

Name: ______________________________ Date: ______________

Unit 2 Practice 16: I can solve story problems with sums of 100.

1) On Monday, Gabriel was working with a total of 100 triangles and squares in the math center. There were 63 triangles and the rest were squares. How many squares was Gabriel working with on Monday?

Gabriel was working with _______ squares on Monday.

2.NBT.5

2) On Tuesday, Gabriel was still working with 100 shapes in the math center. This time he had 31 triangles, 27 squares and the rest were circles. How many circles was Gabriel working with on Tuesday?

Gabriel was working with _______ circles on Tuesday.

2.NBT.5

Do you understand? ? ✓

Name: ______________________________ Date: ______________

Unit 2 Practice 16: I can solve story problems with sums of 100.

3) On Wednesday, Gabriel was still working with 100 shapes in the math center. This time he had rectangles, circles and hexagons. What is one combination of rectangles, circles and hexagons Gabriel could have?

Gabriel could have _______ rectangles, _______ circles and _______ hexagons

2.NBT.5

4) On Thursday, Gabriel had triangles and circles to work with in the math center. He had 42 triangles and 16 more circles than triangles to work with in the math center. Gabriel told his teacher that he was working with a total of 100 shapes. Do you agree or disagree with Gabriel?

I _________________ with Gabriel.

2.NBT.5

Do you understand? **?** ✓

Name: ____________________ Date: __________

I can practice grade level fluencies.

Set 1: Solve.	**Set 2:** Find the **missing value.**
1) 10 + 4 = _______	1) 10 + _______ = 20
2) 9 + 5 = _______	2) 6 + _______ = 10
3) 20 - 4 = _______	3) 7 + _______ = 10
4) 18 - 2 = _______	4) 18 = 2 + _______
5) 17 + 3 = _______	5) 20 = 9 + _______
6) 17 + 4 = _______	6) 8 + _______ = 13
7) 18 - 9 = _______	7) 20 = 8 + _______
2.NBT.5	2.NBT.5

Do you understand? ? ✓

Unit Reflection

Adding and Subtracting within 100

Use this space to reflect on your understanding of the unit skills and concepts.

Skill/Concept	I can...	I need to work on...

Unit 3

Measuring Length

In this unit, we will measure how long things are using centimeters, meters, inches, and feet. We will use our measurements to solve story problems and show data on a line plot.

Measuring Length using Centimeters and Using Inches

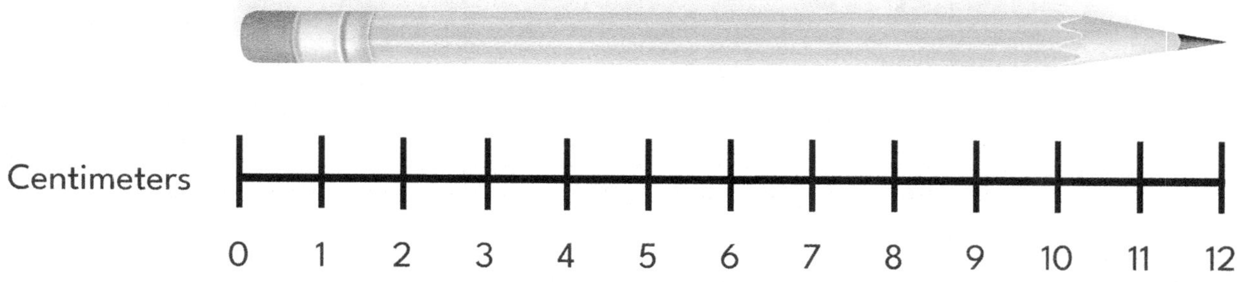

The pencil is 12 centimeters long.

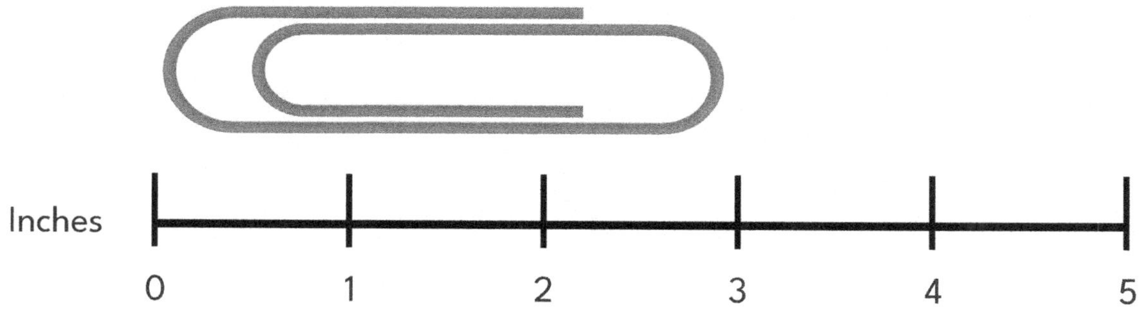

The paperclip is 3 inches long.

Unit Vocabulary

Measuring Length

Use this space to visualize the math vocabulary for this unit.

Word or Phrase	Example or Attributes	Visual Reminder

Unit Models & Strategies

Measuring Length

Use this space to visualize the math models and strategies for this unit.

Model or Strategy	This is a...	It is used to...

Name: ______________________________ Date: ______________

Unit 3 Practice 1: I can use cubes to measure objects accurately.

1) How many cubes long is each pencil?

Pencil A

Pencil B

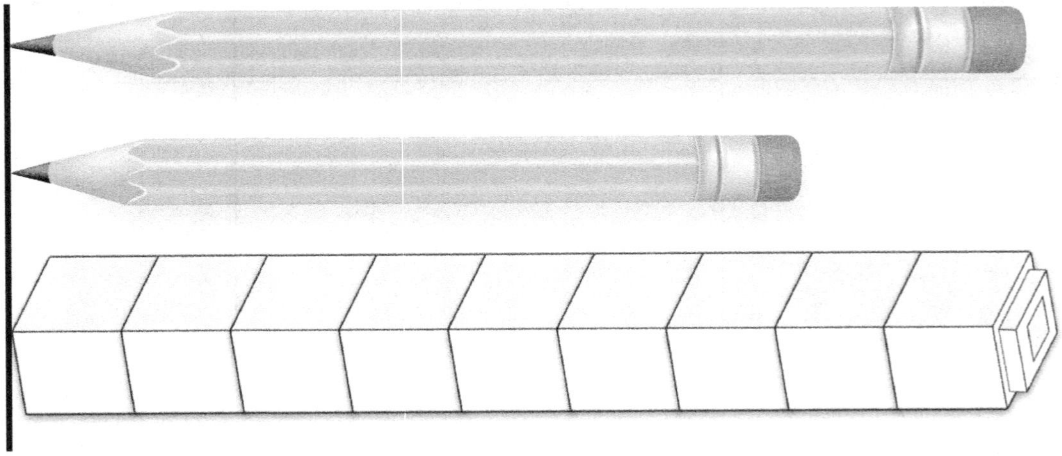

Pencil A is _______ cubes long.   Pencil B is _______ cubes long.

1.MD.2

2) How many cubes long is each string?

String A

String B

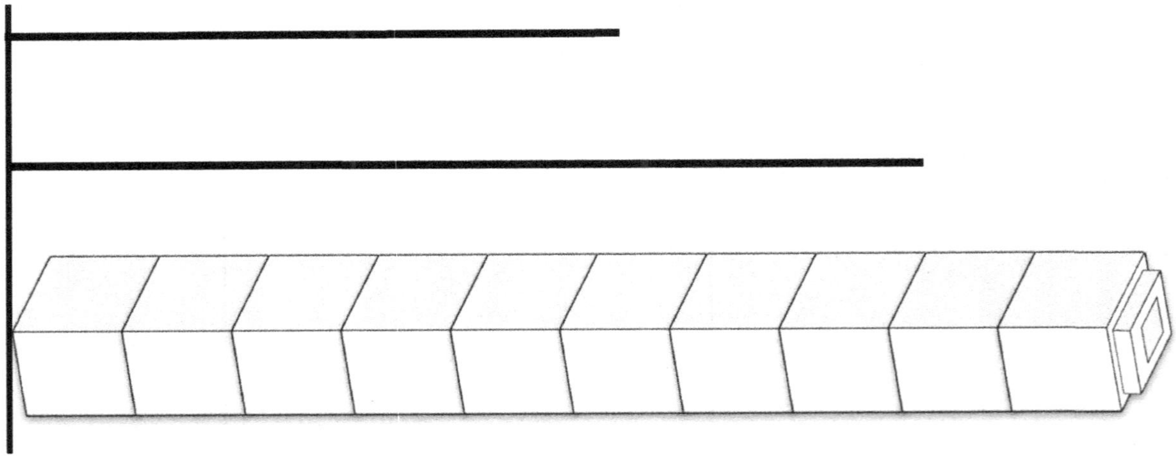

String A is _______ cubes long.   String B is _______ cubes long.

1.MD.2

Do you understand? **?** ✓

Name: ______________________________ Date: ______________

Unit 3 Practice 1: I can use cubes to measure objects accurately.

3) **Agree or Disagree**:
Ellie measured a string using cubes. She says that the string is 5 cubes long. Do you agree or disagree with Ellie?

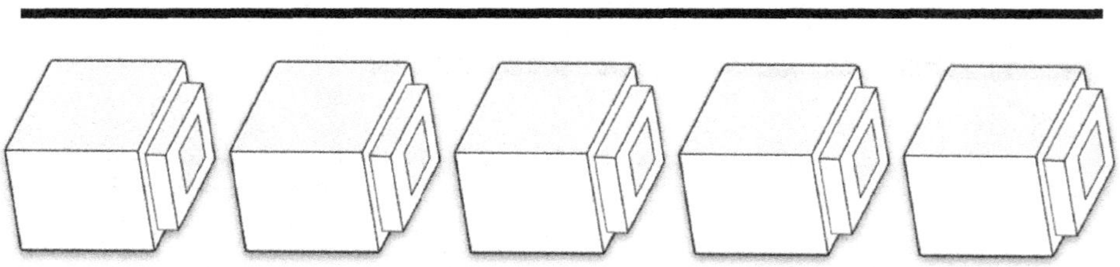

I _______________ with Ellie.

1.MD.2

4) **Agree or Disagree**:
Owen measured a paperclip using cubes. He says that the paper clip is 7 cubes long. Do you agree or disagree with Owen?

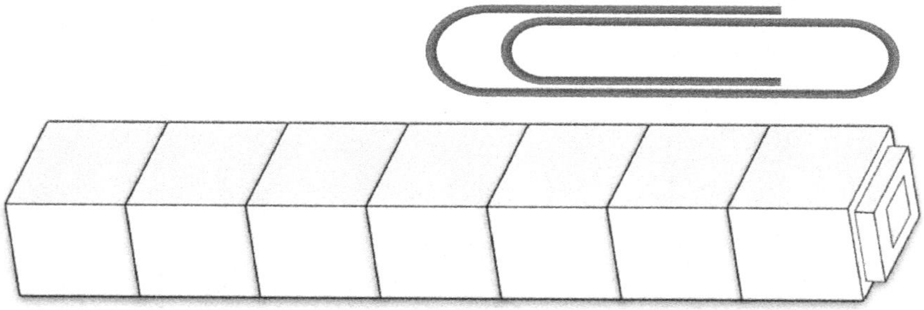

I _______________ with Owen.

1.MD.2

Do you understand? **?** ✓

Affirmation: Every day that I keep trying, is one day closer to me getting it.

Name: ______________________ Date: ____________

Unit 3 Practice 2: I can use base-ten block cubes to measure objects.

1) How many base-ten block unit cubes long is each ribbon?

Ribbon A

Ribbon B

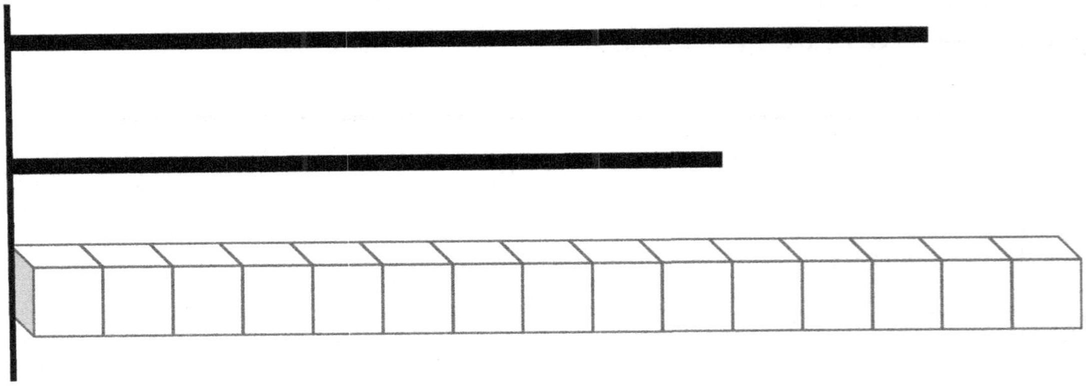

Ribbon A is _______ base-ten cubes long.

Ribbon B is _______ base-ten cubes long.

1.MD.2

2) How many base-ten block unit cubes long is each paperclip?

Paperclip A

Paperclip B

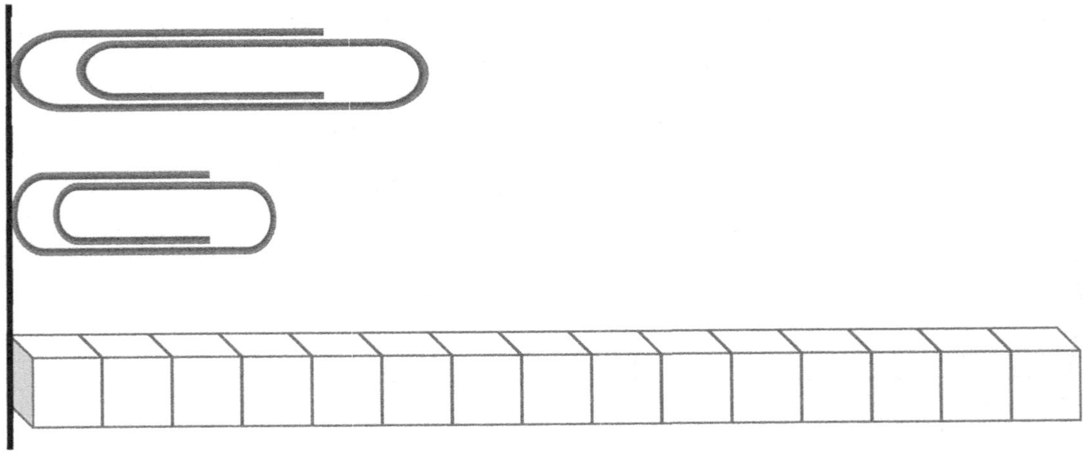

Paperclip A is _______ base-ten cubes long.

Paperclip B is _______ base-ten cubes long.

1.MD.2

Do you understand? ? ✓

Affirmation: Every day that I keep trying, is one day closer to me getting it.

Name: ______________________ Date: __________

Unit 3 Practice 2: I can use base-ten block cubes to measure objects.

3) Draw a pencil that is longer than 10 base-ten block cubes.

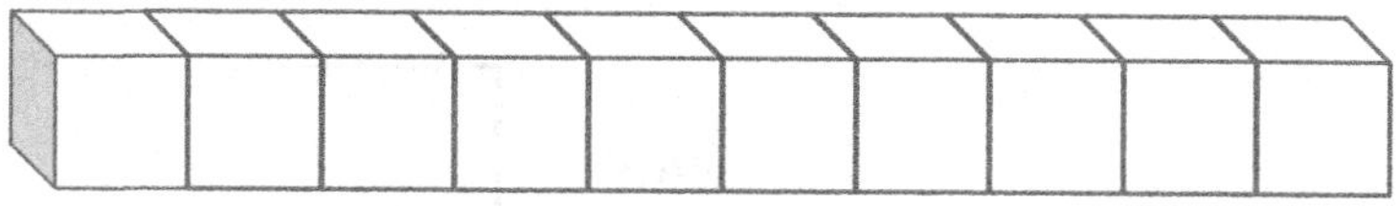

My pencil is ______ base-ten block cubes long.

2.MD.1

4) Draw a crayon that is shorter than 10 base-ten block cubes.

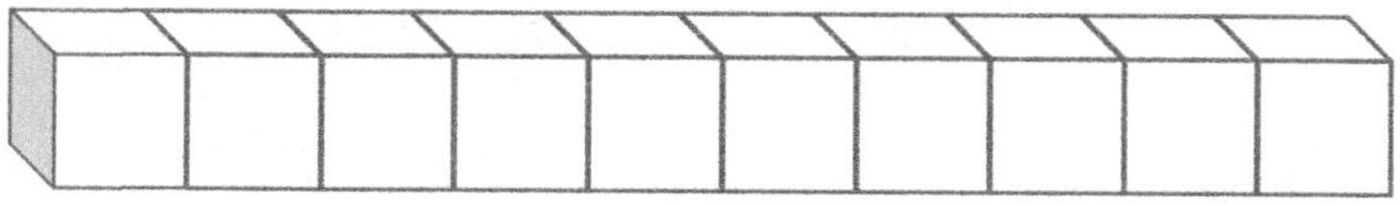

My crayon is ______ base-ten block cubes long.

2.MD.1

Do you understand? **?** ✓

Name: ______________________________ Date: ______________

Unit 3 Practice 3: I can make a model of a centimeter ruler to measure objects.

1) Solve

Find the **sum**	Find the **difference**
25 + 20 = ☐	25 - 20 = ☐
25 + 24 = ☐	25 - 24 = ☐
37 + 30 = ☐	37 - 20 = ☐
37 + 32 = ☐	37 - 25 = ☐

2.NBT.5

2) Same or different. Which are the same length? Which are different lengths?

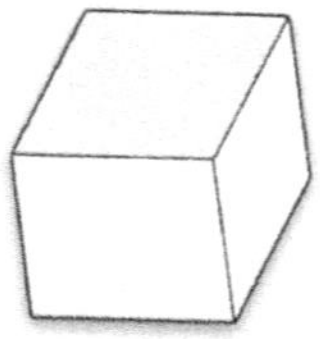

base-ten cube

paperclip

dime

pencil eraser

Same	Different

1.MD.2

Do you understand? **?** ✓

Affirmation: Believing that I can be successful in math means that I can do it!

Name: ______________________________ Date: ______________

Unit 3 Practice 3: I can make a model of a centimeter ruler to measure objects.

3) A pencil eraser is about 1 centimeter long.
Use a pencil eraser to make a model of a centimeter ruler.
Include tick marks and numbers.

1.MD.2

4) Use the model of a centimeter ruler to measure the lines.

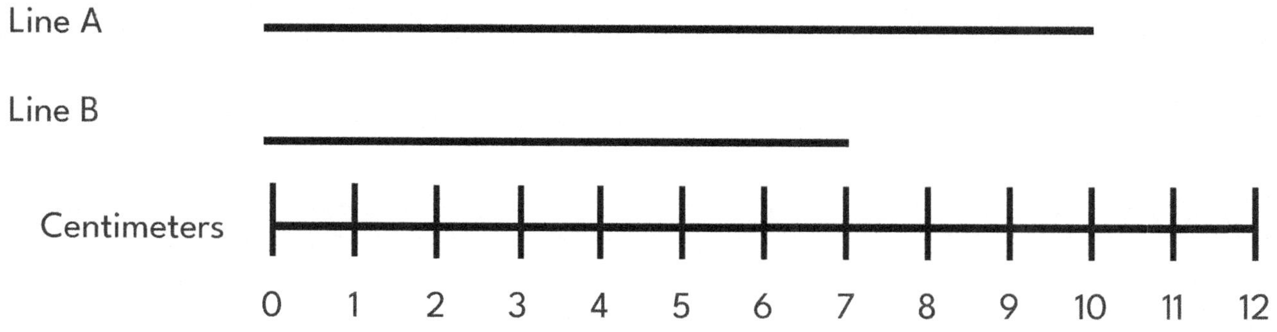

Line A is _______ centimeters long. Line B is _______ centimeters long.

Which line is longer? Line _______ is longer than Line _______.

2.MD.4

Do you understand? **?** ✓

Name: ______________________________ Date: ______________

Unit 3 Practice 4: I can estimate and compare lengths using centimeters.

1) About how long is the marker in centimeters?

The marker is about _______ centimeters long.

2.MD.3

2) About how long is the paperclip in centimeters?

The paperclip is about _______ centimeters long.

2.MD.3

Do you understand? **?** ✓

Name: ______________________________ Date: ______________

Unit 3 Practice 4: I can estimate and compare lengths using centimeters.

3) Paula made two trains using base-ten block cubes.
How much longer is train A than train B?

Train A

Train B

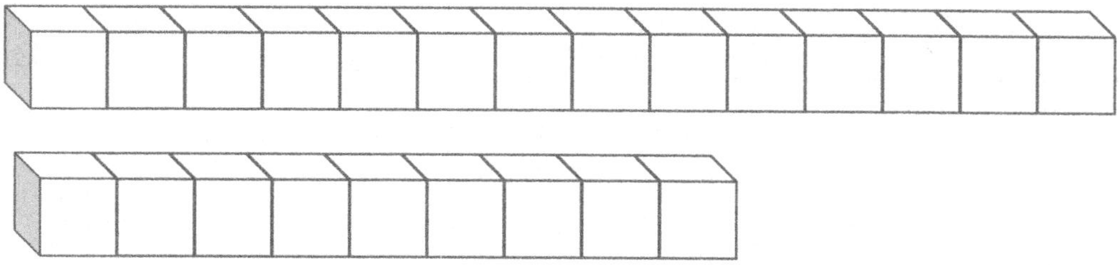

Train A is _______ base-ten blocks longer than train B.

2.MD.4

4) How much longer?

Line A

Line B

Centimeters

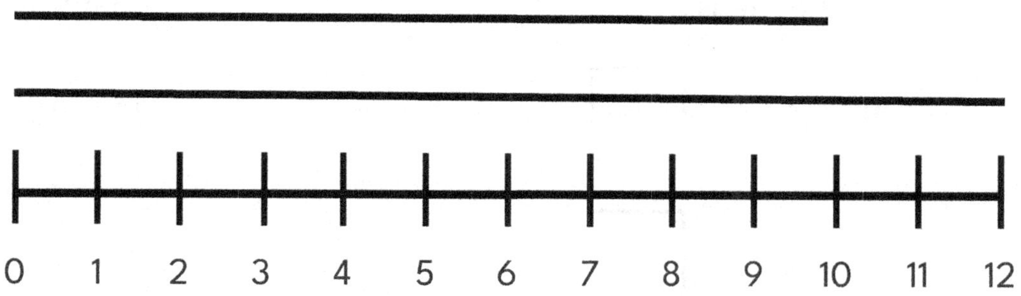

Line A is _______ centimeters long. Line B is _______ centimeters long.

How much longer is Line B than Line A? Line B is _______ centimeters longer.

2.MD.4

Do you understand? ? ✓

Name: ______________________ Date: ____________

Unit 3 Practice 5: I can choose objects to measure using centimeters or meters.

1) Count on by tens.

50, ______, ______, ______, ______, ______, ______, ______, ______, 140

15, ______, ______, ______, ______, ______, ______, ______, ______.

48, ______, ______, ______, ______, ______, ______, ______, ______.

36, ______, ______, ______, ______, ______, ______, ______, ______.

2.NBT.2

2) Solve.

Add tens		Subtract tens	
79 + 10 =	☐	127 - 10 =	☐
79 + 20 =	☐	127 - 20 =	☐
79 + 30 =	☐	127 - 30 =	☐
79 + 40 =	☐	127 - 40 =	☐

2.NBT.2

Do you understand? **?** ✓

Affirmation: Solving problems is what humans do!

Name: ______________________________ Date: ______________

Unit 3 Practice 5: I can choose objects to measure using centimeters or meters.

3) Thomas has a pet cat that is 42 centimeters long. Thomas saw a lion exhibit at the zoo. Thomas wondered if the lion should be measured using centimeters or using meters. What do you think?

I think the lion should be measured using ______________.

2.MD.1

4) Olivia measured the height of her book in centimeters. The height of her book was 28 centimeters. Asher measured the height of the classroom door. He said that the height of the door was 2 centimeters. Olivia thinks that Asher made a mistake. Explain the mistake Asher could have made.

I think...

____

____

____

____

2.MD.1

Do you understand? **?** ✓

Name: ______________________________ Date: ______________

Unit 3 Practice 6: I can solve story problems using unit measures.

1) Count on by fives.

10, 15, ______, ______, ______, ______, ______, ______, ______, 55

75, 80, ______, ______, ______, ______, ______, ______, ______, 120

60, ______, ______, ______, ______, ______, ______, ______, ______, 105

105, ______, ______, ______, ______, ______, ______, ______, 145

2.NBT.2

2) Match the number that makes each equation true.

60 + 5 =	☐	90
75 + 15 =	☐	55
25 + 25 =	☐	65
30 + 25 =	☐	50

2.NBT.2

Do you understand? **?** ✓

Name: ______________________________ Date: ____________

Unit 3 Practice 6: I can solve story problems using unit measures.

Questions A and B refer to the information below.

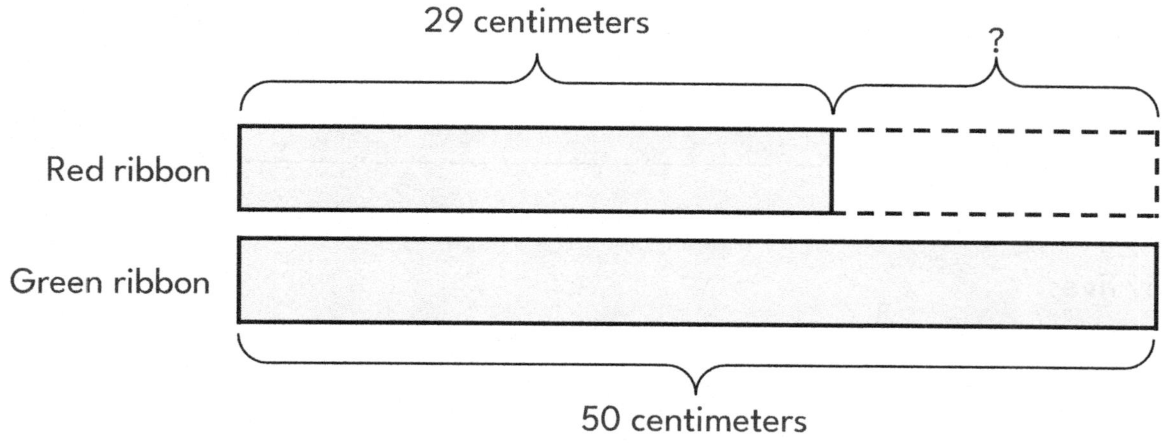

A) The tape diagram shows the length of ribbons Leo cut for an art project.

Leo's red ribbon is ______________________ long.

Leo's green ribbon is ______________________ long.

2.MD.5

B) How much longer is the green ribbon than the red ribbon?

The green ribbon is _______ centimeters longer than the red ribbon.

2.MD.5

Do you understand? **?** ✓

Name: ______________________________ Date: ______________

Unit 3 Practice 7: I can solve story problems using estimation and measurement.

1) Count back by tens.

100, 90, ______, ______, ______, ______, ______, ______, ______, 10.

115, 105, ______, ______, ______, ______, ______, ______, ______, 25.

Count back by fives.

100, 95, ______, ______, ______, ______, ______, ______, ______, 55.

145, 140, ______, ______, ______, ______, ______, ______, ______, 100.

2.MD.2

2) Circle true or false for each equation.

105 - 15 = 100

100 - 35 = 65

90 - 25 = 65

70 - 45 = 35

2.NBT.5

Do you understand? ? ✓

Name: ______________________________ Date: ____________

Unit 3 Practice 7: I can solve story problems using estimation and measurement.

3) Ava estimated the length of her dog as 58 centimeters (this included his tail). She measured her dog and he was 53 centimeters long. How many centimeters shorter was the length of Ava's dog than her estimate?

Ava's dog was _______ centimeters shorter than her estimate.

2.MD.5

4) Saul estimated the length of his desk. His desk measured 79 centimeters long. His estimate was 15 centimeters less than the actual length of his desk. What was Saul's estimate?

Saul's estimate was _______ centimeters.

2.MD.5

Do you understand? **?** ✓

Name: ____________________ Date: __________

Unit 3 Practice 8: I can measure objects using inches.

1) Match the shape to its name.

square rectangle triangle

1.G.1

2) Same or different. Which are about the same length? Which are different lengths?

paperclip

counter

square tile

quarter

Same	Different

1.MD.2

Do you understand? **?** ✓

Affirmation: When I am nervous or scared trying is brave and courageous!

Name: ______________________________ Date: ______________

Unit 3 Practice 8: I can measure objects using inches.

3) A quarter is about 1 inch long.
Use a quarter to make a model of an inch ruler.
Include tick marks and numbers.

1.MD.2

4) The pencil is measured in inches and in centimeters.

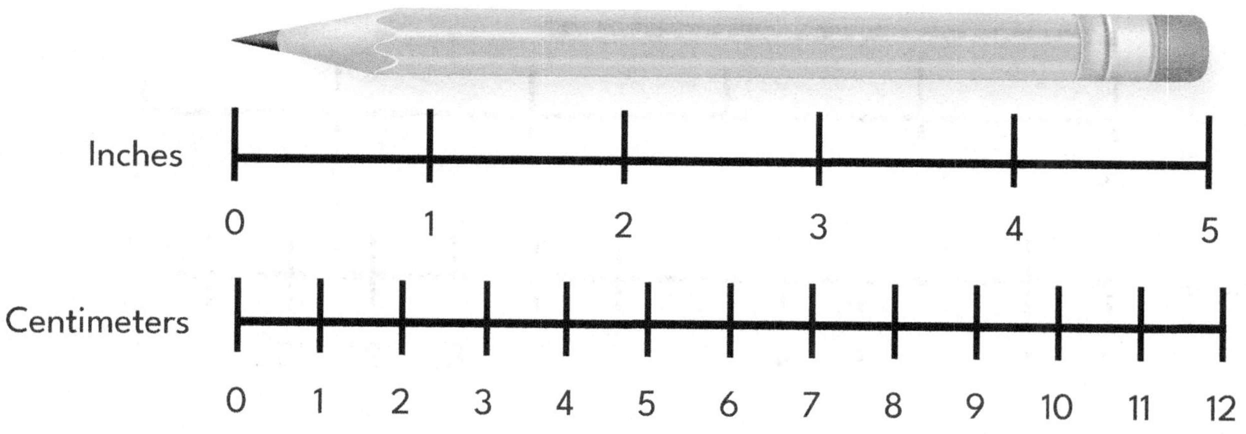

The pencil is about _______ inches. The pencil is about _______ centimeters.

2.MD.2

Do you understand? **?** ✓

Affirmation: I don't just memorize math facts! I understand math facts!

Name: ______________________________ Date: ______________

Unit 3 Practice 9: I can compare smaller and larger units.

1) Find the missing numbers.

12 = 6 + ______	8 + 4 = ______	24 = 12 + 10 + ______
______ = 7 + 5	12 + 6 = ______	18 = 10 + 2 + ______
12 = ______ + 2	12 + 12 = ______	8 + 4 + ______ = 24

2.OA.2

2) Luna explained that that the line measured about 3 inches or about 7 centimeters.

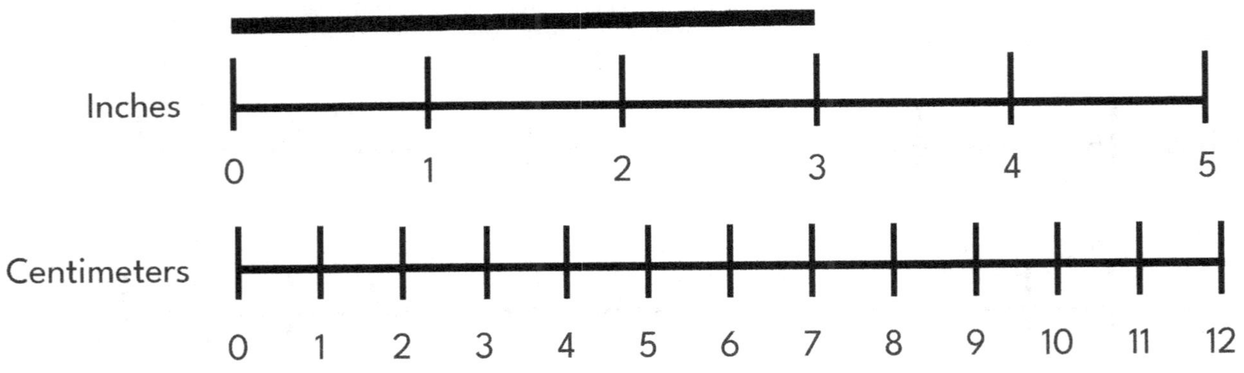

Explain why measuring with inches is different from measuring with centimeters.

2.MD.2

Do you understand? **?** ✓

Name: ______________________________ Date: ______________

Unit 3 Practice 9: I can compare smaller and larger units.

3) Measure the rectangle in inches.

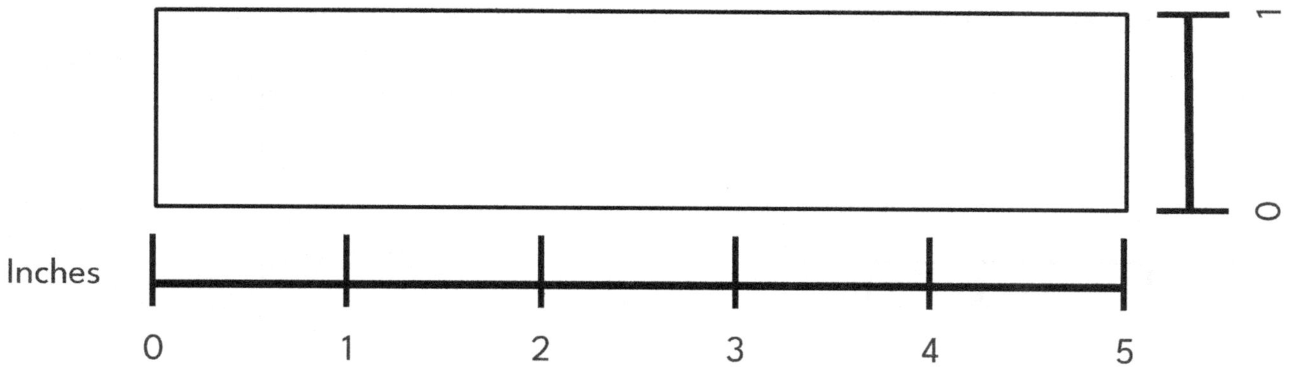

The long side of the rectangle is _______ inches long.

The short side of the rectangle is _______ inches long.

2.MD.1

4) **Agree or Disagree**:
Peter's pet snake is 1 foot long. Milo says that Peter's pet snake is 12 inches long. Do you agree or disagree with Milo?

I __________________ with Milo.

2.MD.2

Do you understand? **?** ✓

Affirmation: I learn from the ideas of other people, even if I don't agree with them.

Name: ______________________ Date: ____________

Unit 3 Practice 10: I can measure objects with a broken ruler.

1) Match the expressions that are equal.

9 + 3	6 + 4
12 + 2	14 - 2
18 - 8	10 - 5
8 - 3	10 + 4

1.OA.3

2) Draw a crayon that is 4 inches long.

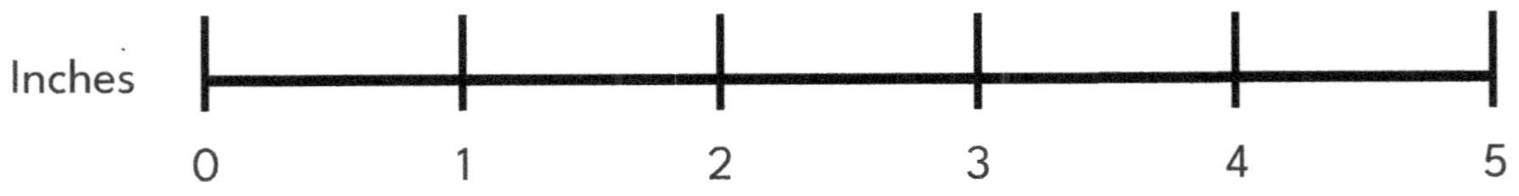

Explain how you used the ruler to draw the 4 inch crayon.

2.MD.6

Do you understand? **?** ✓

Affirmation: I learn from the ideas of other people, even if I don't agree with them.

Name: ______________________________ Date: ______________

Unit 3 Practice 10: I can measure objects with a broken ruler.

Questions A and B refer to the information below.

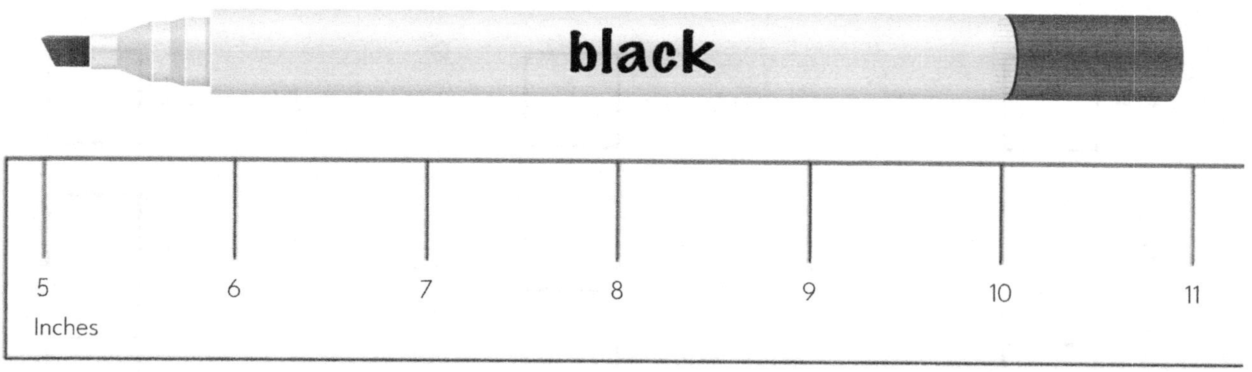

A) **Agree or Disagree**:
Henry says that the marker is 11 inches long. Do you agree or disagree with Henry?

I ______________ with Henry.

2.MD.5

B) What is the length of the marker in inches? Use drawings, numbers or words to show your thinking.

The marker is _______ inches long.

2.MD.5

Do you understand? **?** ✓

Affirmation: I encourage myself and others by saying kind words.

Name: ______________________________ Date: ______________

Unit 3 Practice 11: I can solve one-step story problems using length.

1) Add and subtract within 50.

15 + 5 = ☐	24 - 4 = ☐	50 - 25 = ☐
20 + 12 = ☐	24 - 5 = ☐	50 - 26 = ☐
12 + 8 = ☐	12 - 12 = ☐	34 + 6 = ☐

2.NBT.5

2) Brooke has a rope that is 36 inches long. She cut 12 inches off the rope for an art project. How long is the remaining rope?
Use the tape diagram to solve.

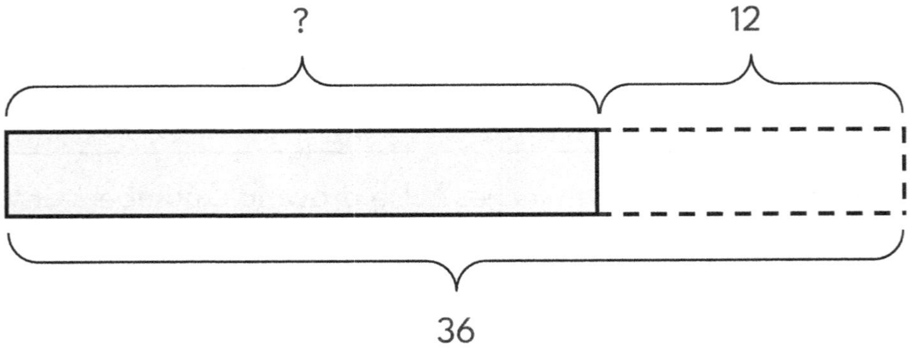

Equation: ______________________________

Brooke has _______ inches of rope remaining.

2.MD.5

Do you understand? **?** ✓

Affirmation: I encourage myself and others by saying kind words.

Name: ______________________________ Date: ______________

Unit 3 Practice 11: I can solve one-step story problems using length.

3) Felix started with a piece of red yarn. He cut off 26 inches of the red yarn for an art project and now has 24 inches left. How much red yarn did Felix have to start? Construct a tape diagram to solve.

Equation: ______________________________

Felix started with _______ inches of red yarn.

2.MD.5

4) Mia needs 8 feet of rope to use as a jump rope. If the supply room has a rope that is 37 feet long, how much rope is left after Mia cuts off the piece she needs?

Equation: ______________________________

There is _______ feet of rope left in the supply room.

2.MD.5

Do you understand? **?** ✓

Name: ______________________________ Date: ______________

Unit 3 Practice 12: I can solve two-step story problems using length.

1) Find the sums.

30 + 20 + 7 = ☐ 45 + 22 + 8 = ☐

6 + 6 + 12 = ☐ 67 + 3 + 14 = ☐

7 + 4 + 8 = ☐ 52 + 24 + 24 = ☐

2.NBT.5

2) Ben is using three different colors of yarn for book binding. He has 23 inches of green yarn, 14 inches of red yarn and 31 inches of yellow yarn. How much yarn, in inches, does Ben have in total to use for book binding?

Ben has _______ inches of yarn.

2.MD.5

3) Ann drew a triangle with the side lengths of 6 inches, 6 inches and 5 inches. How much red yarn will she need to outline her triangle with red yarn?

Ann will need _______ inches of red yarn to outline her triangle.

2.MD.5

Do you understand? **?** ✓

Name: ______________________________ Date: ______________

Unit 3 Practice 12: I can solve two-step story problems using length.

4) Find the difference.

30 - 20 = ☐

100 - 82 = ☐

20 - 6 = ☐

82 - 12 = ☐

14 - 12 = ☐

70 - 35 = ☐

2.NBT.5

5) The supply room has 90 feet of rope. Mr. Lopez cut off two pieces of rope to make jump ropes for recess. The first jump rope is 12 feet long and the second jump rope is 8 feet long. How much rope is in the supply room now?

There is _______ feet of rope left in the supply room.

2.MD.5

6) Pam had a piece of yarn that was 65 inches long. She cut off some yarn for her brother and some more yarn for her sister. Pam now has 25 inches of yarn left. How much yarn, in inches, could she have given to her brother and to her sister?

Pam could have given ______ inches of yarn to her brother and ______ inches of yarn to her sister.

2.OA.1b

Do you understand? **?** ✓

Name: ______________________________ Date: ______________

Unit 3 Practice 13: I can solve problems using estimation and measurement.

1) Caleb had a piece of green yarn that was 47 inches long. He cut off a piece that he **estimated** was 25 inches long and gave it to his sister. When he measured the yarn he had left, he had 35 inches of yarn left. How much yarn did Caleb give to his sister?

Caleb gave _______ inches of green yarn to his sister.
Circle one, Caleb's **estimate** was:

too high / too low / about right.

2.MD.3

2) **Agree or Disagree**:
Aria was playing *Target Measurement*. The target length was 1 foot. She drew a line that was 11 inches long. Aria said that she should get 1 point for being one inch from the target length, the difference between 1 foot and 11 inches. Do you agree or disagree with Aria?

I ________________ with Aria.

2.MD.3

Do you understand? **?** ✓

Affirmation: By helping others, I help myself learn more.

Name: ______________________________ Date: ______________

Unit 3 Practice 13: I can solve problems using estimation and measurement.

Questions A and B refer to the information below.

A) A quarter is about one inch. **About** how long is each side of the rectangle in inches?

My **estimate**:

The long side of the rectangle is is about _______ inches long.

The short side of the rectangle is about ______ inches long.

2.MD.3

B) Use a quarter to measure each side of the rectangle.

Using a quarter, the long side of the rectangle is ______ units long. The short side of the rectangle is ______ units long.

Circle one, my estimates were:

too high / too low / about right.

2.MD.3

Do you understand? **?** ✓

Name: ______________________________ Date: ______________

Unit 3 Practice 14: I can answer questions about data in a line plot.

1) Mason measured these tools from the art center.

pencil

paintbrush

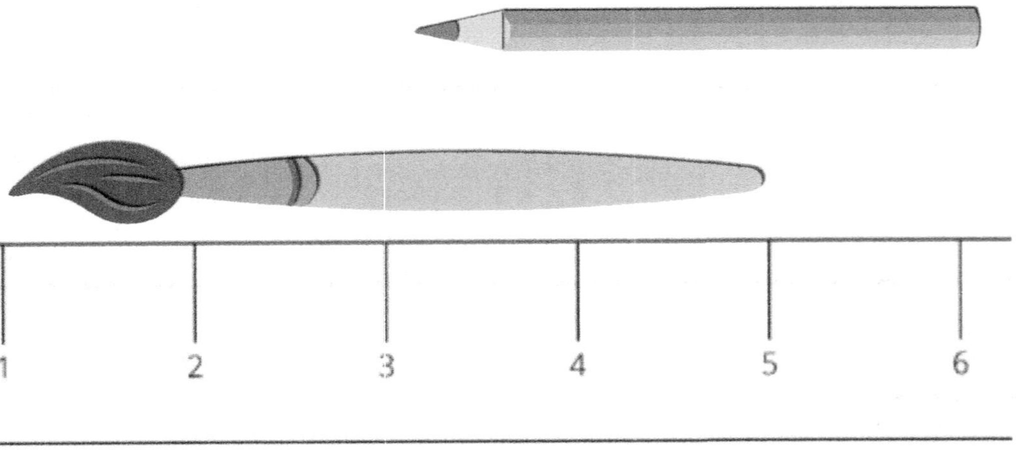

The ____________ is the longest art tool. The ____________ is the shortest art tool.

2.MD.4

2) A pet store has a tank with 9 goldfish. The line plot shows the length of each goldfish.

Length of Goldfish

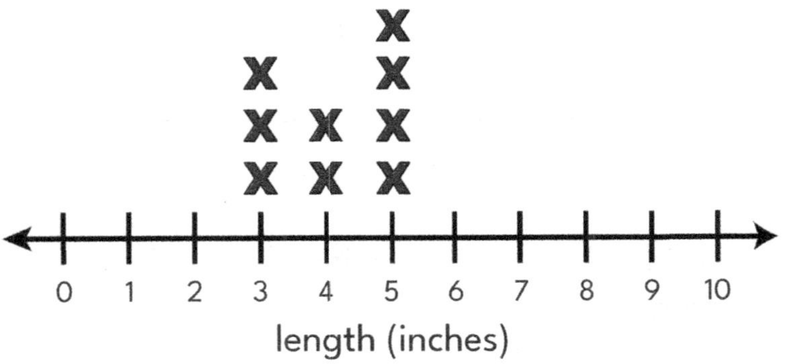

The longest goldfish is _____ inches long. The shortest goldfish is _____ inches long.

2.MD.9

Do you understand? **?** **✓**

Name: ______________________________ Date: ______________

Unit 3 Practice 14: I can answer questions about data in a line plot.

Questions A and B refer to the graph below.

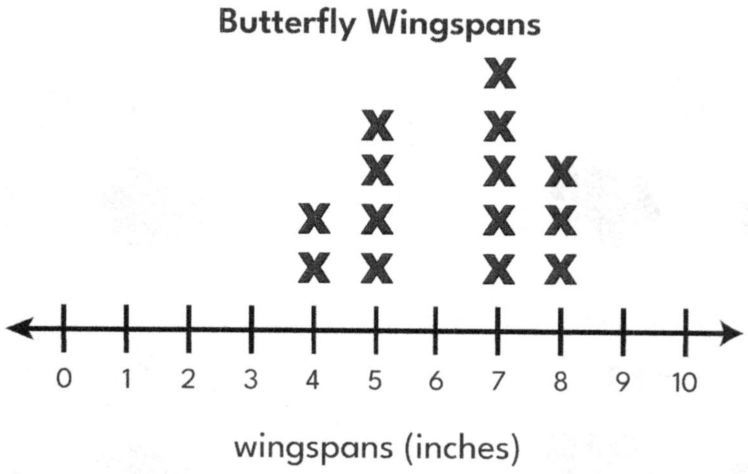

A) How many butterflies have a wingspan of 4 inches?

_______ butterflies have a wingspan of 4 inches.

How many butterflies have a wingspan of 7 inches?

_______ butterflies have a wingspan of 7 inches.

2.MD.9

B) How many butterfly wingspans were measured for the line plot?

_______ butterfly wingspans were measured for the line plot.

2.MD.9

Do you understand? **?** ✓

Name: ______________________________ Date: ______________

Unit 3 Practice 15: I can show data in a line plot.

1) Circle true or false for each equation.

52 - 30 = 20

58 - 25 = 33

85 - 35 = 50

58 - 15 = 33

2.NBT.5

2) Draw a line that is **longer** than 10 centimeters.
Draw a line that is **shorte**r than 10 centimeters.
Label your work.

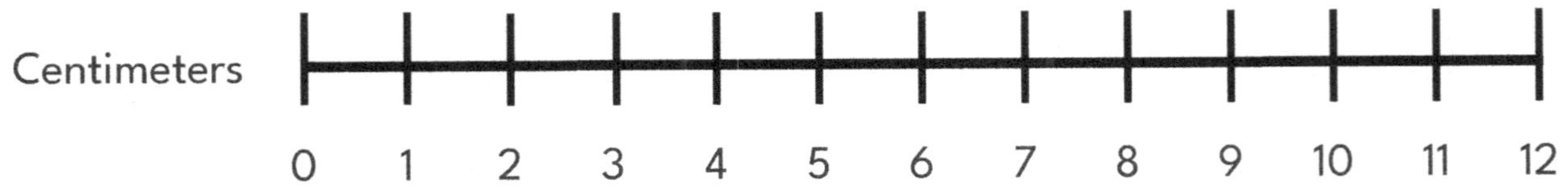

2.MD.4

Do you understand? **?** ✓

Name: ______________________________ Date: ______________

Unit 3 Practice 15: I can show data in a line plot.

Questions A and B refer to the graph below.

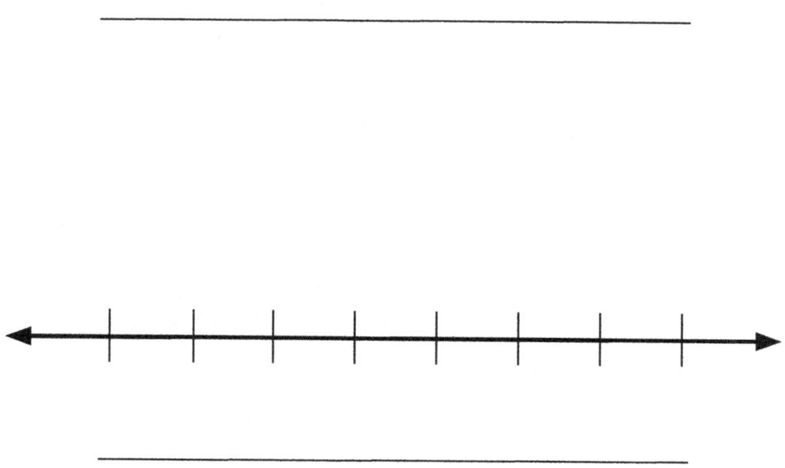

Grasshopper Length (centimeters)
7
3
4
3
5
7

A) Use the data in the table to create a line plot.

Remember to:

- ❏ give the line plot a title
- ❏ use Xs to show each measure
- ❏ label the scale (including measurement unit)
- ❏ write numbers for each interval

2.MD.9

B) How many grasshoppers are shown in this data?

_______ grasshopper lengths are shown in this data.

2.MD.9

Do you understand? **?** ✓

Name: ______________________________ Date: ______________

Unit 3 Practice 16: I can make a line plot using data from a table.

1) Write the number that makes each equation true.

63 + 10 + _______ = 83

27 + _______ + 2 = 30

79 = 39 + _______

53 = 25 + 25 + _______

10 + 10 + _______ = 45

8 + 8 + _______ = 20

19 + 19 + 19 = _______

11 + 12 + 13 = _______

2.NBT.5

2) Mary has a collection of 42 buttons that measure one centimeter, two centimeters or three centimeters. **Most** of the buttons measure 2 centimeters. The **fewest** number of buttons measure 1 centimeter. The rest of the buttons in her collection measure 3 centimeters. How many of each size button could she have in her collection?

Mary could have _______ one centimeter buttons, _______ two centimeter buttons and _______ three centimeter buttons.

2.OA.1

Do you understand? ? ✓

Affirmation: What matters is that I tried, even if I was the last one to finish.

Name: ______________________ Date: ____________

Unit 3 Practice 16: I can make a line plot using data from a table.

3) Frogs can jump ten times the length of their bodies. The table shows the distance the frogs at a state fair jumped during a frog jumping contest.
Create a line plot showing this data.

Frog Contestants	Length of jump (inches)
Fred	50
Beth	50
Amy	55
Chris	53
Harry	53
Frank	54
Susan	54
Omar	55
Ben	50
Carmen	50
Theo	53
Carl	53

____

____

2.MD.9

Do you understand? **?** ✓

Name: ______________________________ Date: ______________

Unit 3 Practice 17: I can answer questions about data in a line plot.

1) Find the difference.

90 - 20 = ☐	80 - 11 = ☐
90 - 21 = ☐	80 - 13 = ☐
90 - 22 = ☐	40 - 21 = ☐
90 - 23 = ☐	40 - 24 = ☐
90 - 25 = ☐	50 - 24 = ☐

2.NBT.5

2) Cyrus is playing *Target Number*. The target number is 40.
His number cards are 46, 48, 47, 32, 38, 41 and 44.

Which of his number cards are **less than** 40?

His number cards ______________________ are less than 40.

Which of his number cards are **greater than** 40?

His number cards ______________________ are greater than 40.

2.NBT.4

Do you understand? **?** ✓

Name: ______________________________ Date: ______________

Unit 3 Practice 17: I can answer questions about data in a line plot.

Questions A and B refer to the information below.

Garter snakes can grow to 32 inches. The line plot shows the lengths of the garter snakes in a reptile exhibit.

Garter Snake Lengths

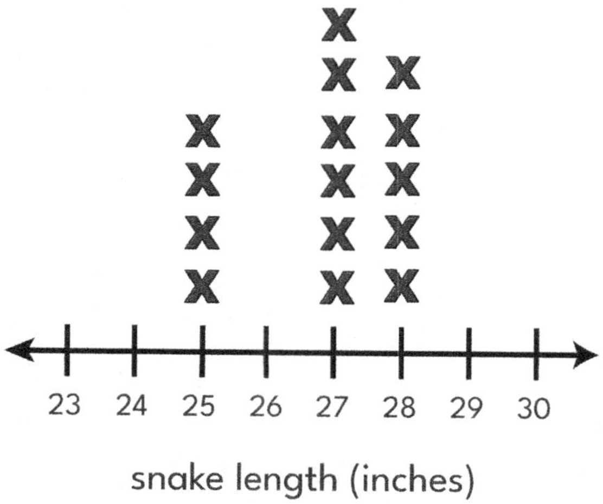

A) How many garter snakes are in the reptile exhibit?

There are _______ garter snakes in the reptile exhibit.

How many snakes measure **less than** 26 inches? _______ sakes measure less than 26 inches.

How many snake measure 27 inches **or more**? _______ snakes measure 27 inches or more.

2.MD.9

B) The longest snake in the exhibit is _______ inches long.

The shortest snake in the exhibit is ______ inches long.

What is the difference in length between the longest and shortest snake in this exhibit?

The difference in length is _______ inches.

2.MD.9

Do you understand? **?** ✓

Name: ______________________________ Date: ______________

Unit 3 Practice 18: I can use inches, feet and yards in a line plot.

1) Match the number that makes each equation true.

$12 + 12 + 12 =$	☐	9
$30 + 30 + 30 =$	☐	36
$39 - 30 =$	☐	31
$12 + 9 + 10 =$	☐	90

2.OA.1b

2) **Agree or Disagree**:
Ezra made a model of a yardstick that showed 36 inches. Violet made a yardstick that showed 3 feet. Violet said that their yardsticks are the same length. Ezra disagreed, he said that his yardstick had numbers 0-36 so his yardstick is longer. Who do you agree with?

I agree with ________________.

1.MD.2

Do you understand? **?** ✓

Name: ______________________________ Date: ______________

Unit 3 Practice 18: I can use inches, feet and yards in a line plot.

Questions A and B refer to the information below.

A gardener has several trees for sale in his tree lot. The trees measure 6 feet to 15 feet tall. The line plot shows the trees measured in yards.

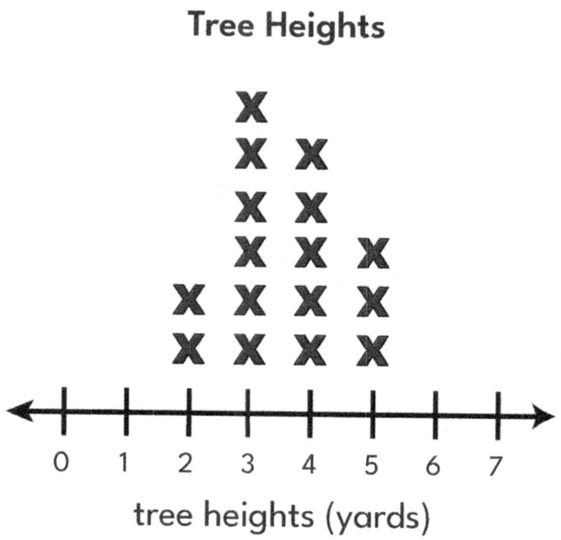

A) How many of the trees are 3 yards or taller?

_______ of the trees are 3 yards or taller.

How many of the trees are 6 yards tall?

_______ of the trees are 6 yards tall.

2.MD.9

B) **Agree or Disagree**:
Jack told Amy that the line plot shows the tallest tree in the lot is 3 yards tall. Amy told Jack that she thinks the tallest tree in the lot is 5 yards tall. Who do you agree with? Use drawings, numbers or words to show your thinking.

I agree with ________________.

2.MD.9

Do you understand? **?** ✓

Name: ______________________________ Date: ______________

I can practice grade level fluencies.

Set 1: Add ten	**Set 2:** Subtract ten
1) 36 + 10 = _______	1) 64 - 10 = _______
2) 36 + 20 = _______	2) 64 - 20 = _______
3) 36 + 30 = _______	3) 64 - 30 = _______
4) 47 + 10 = _______	4) 35 - 10 = _______
5) 57 + 10 = _______	5) 91 - 10 = _______
6) 18 + 10 = _______	6) 89 - 10 = _______
7) 18 + 20 = _______	7) 89 - 20 = _______
2.NBT.8	2.NBT.8

Do you understand? **?** ✓

Unit Reflection

Measuring Length

Use this space to reflect on your understanding of the unit skills and concepts.

Skill/Concept	I can...	I need to work on...

Unit 4

Addition and Subtraction on the Number Line

In this unit, we will learn how a number line works and how to show numbers up to 100 on it. We will use the number line to help us understand and solve addition and subtraction problems.

Number Line

Using a Number Line for Addition

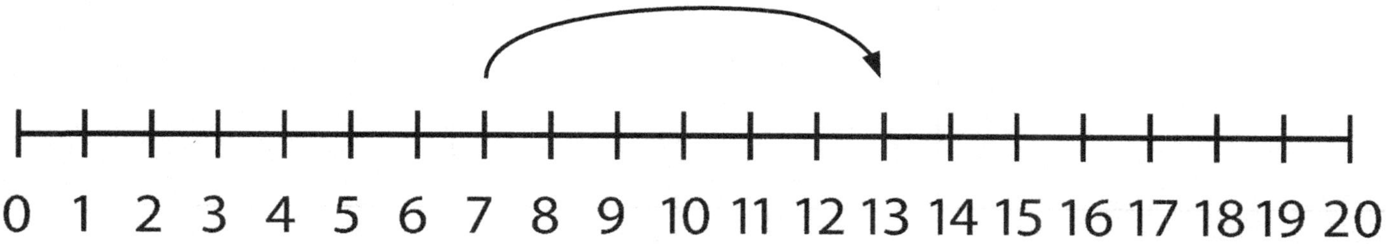

7 + 6 = 13

Using a Number Line for Subtraction

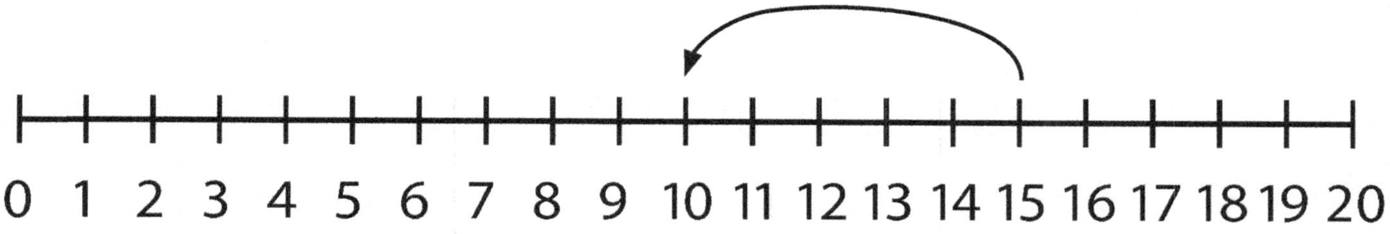

15 - 5 = 10

Unit Vocabulary

Addition and Subtraction on the Number Line

Use this space to visualize the math vocabulary for this unit.

Word or Phrase	Example or Attributes	Visual Reminder

Unit Models & Strategies

Addition and Subtraction on the Number Line

Use this space to visualize the math models and strategies for this unit.

Model or Strategy	This is a...	It is used to...

Name: ______________________________ Date: ______________

Unit 4 Practice 1: I can show numbers on a number line.

1) Count on by ones.

0, 1, ____, ____, ____, ____, ____, ____, ____, ____, 10

5, 6, ____, ____, ____, ____, ____, ____, ____, 14, 15

15, 16, ____, ____, ____, ____, ____, ____, ____, 24, 25

1.NBT.1

2) Fill in the missing numbers on the number line.

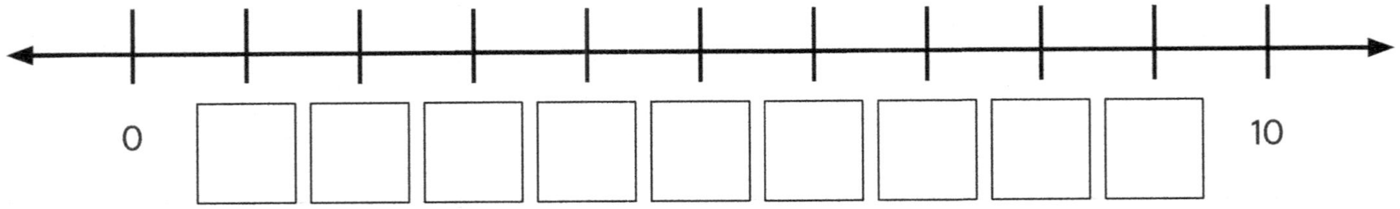

2.MD.6

3) Mark 8 on the number line with a point.

2.MD.6

Do you understand? ? ✓

Affirmation: I am a proficient problem solver!

Name: ______________________________ Date: ______________

Unit 4 Practice 1: I can show numbers on a number line.

4) Label numbers 0 - 12 on the number line.

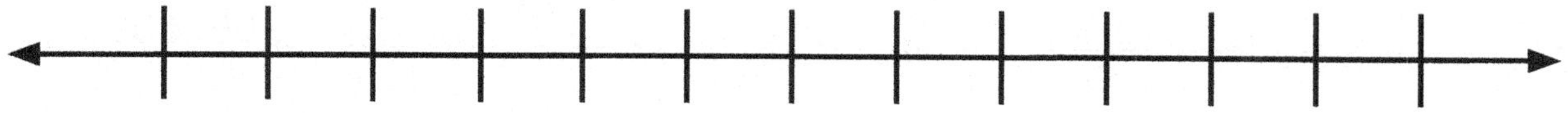

Mark 10 with a point.

2.MD.6

5) **Agree or Disagree**:

Jasper says that an inch ruler is a model of a number line. Emma says that a centimeter ruler is a model of a number line.

Who do you agree with?

I agree with ______________________.

2.MD.6

Do you understand? **?** ✓

Affirmation: My ideas are worthy of sharing.

Name: ______________________________ Date: ______________

Unit 4 Practice 2: I can draw and mark numbers on a number line.

1) Count on by fives.

5, 10, ______, ______, ______, ______, ______, ______, ______, 50

10, 15, ______, ______, ______, ______, ______, ______, ______, 55, 60

45, ______, ______, ______, ______, ______, ______, ______, ______, 90

2.NBT.2

2) Find the sum.

15 + 5 = ______ 5 + 20 = ______

25 + 5 = ______ 5 + 5 + 5 = ______

15 + ______ = 30 20 + 10 + 5 = ______

2.NBT.2

3) Label the missing numbers on the number line.

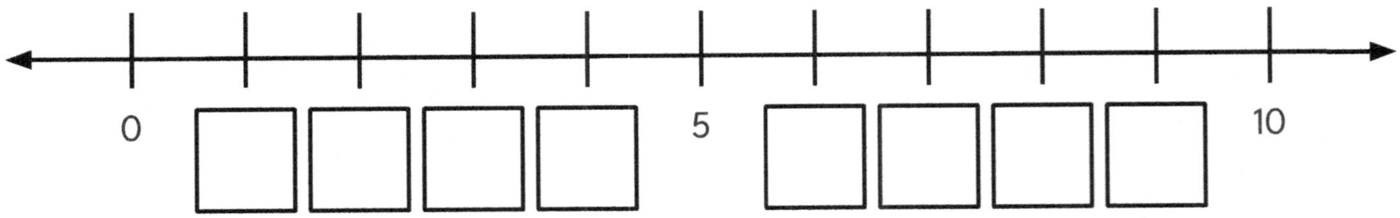

Explain how you found the missing numbers.

2.NBT.2

Do you understand? **?** ✓

Name: ______________________________ Date: ____________

Unit 4 Practice 2: I can draw and mark numbers on a number line.

4) Circle true or false to show if the number line is accurate.

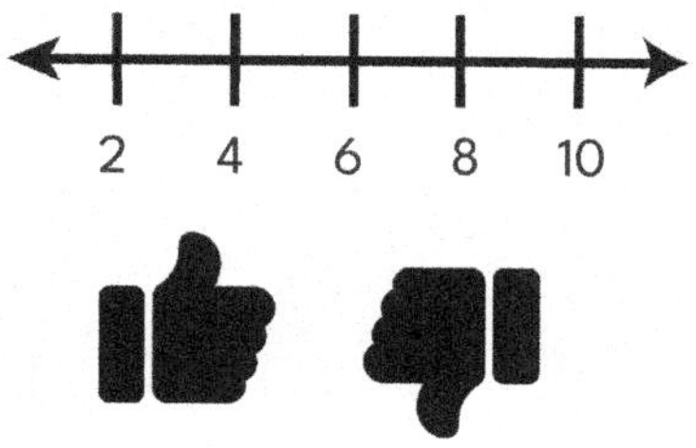

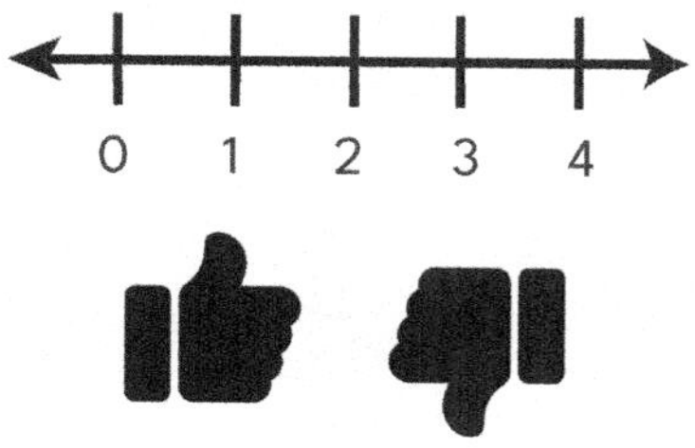

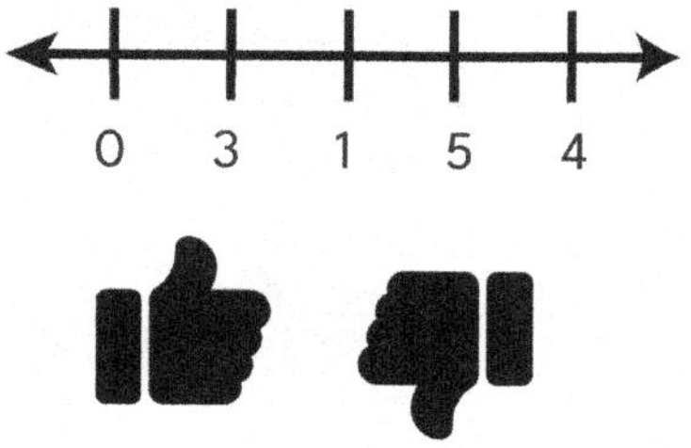

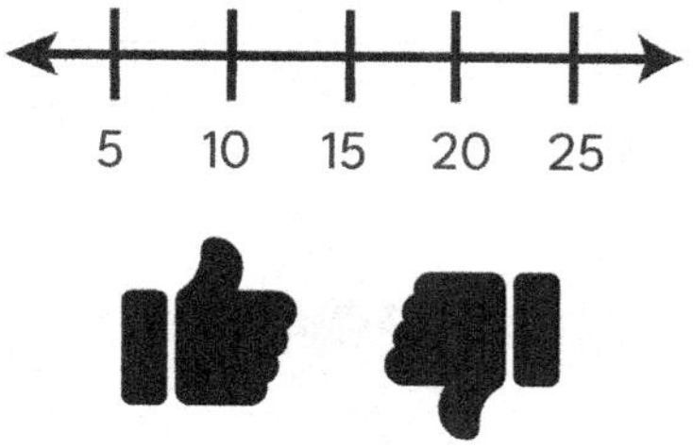

2.MD.6

5) Draw a number line showing numbers 0 - 10.

Mark 3 on your number line with a point.

Mark 7 on your number line with a point.

Which point is closer to zero? _______ is closer to zero.

2.MD.6

Do you understand? ? ✓

Name: ______________________________ Date: ______________

Unit 4 Practice 3: I can find missing numbers on a number line.

1) Count on by tens.

0, 10, ______, ______, ______, ______, ______, ______, ______, 90

5, 15, ______, ______, ______, ______, ______, ______, ______, 95

32, 42, ______, ______, ______, ______, ______, ______, ______, 122

2.NBT.2

2) Count on by fives.

4, 9, ______, ______, ______, ______, ______, ______, ______, 49

6, 11, ______, ______, ______, ______, ______, ______, ______, 51

22, 27, ______, ______, ______, ______, ______, ______, ______, 67

2.NBT.2

3) What's in between? Fill in the missing numbers on the number line.

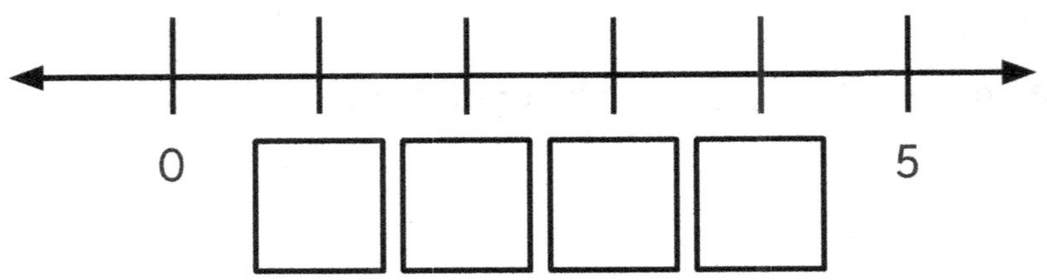

2.MD.9

Do you understand? **?** ✓

Affirmation: Everything that I have learned was something that I did not know at some point.

Name: ____________________ Date: __________

Unit 4 Practice 3: I can find missing numbers on a number line.

4) Draw tick marks on the number line to show the numbers that are not labeled counting on by ones.

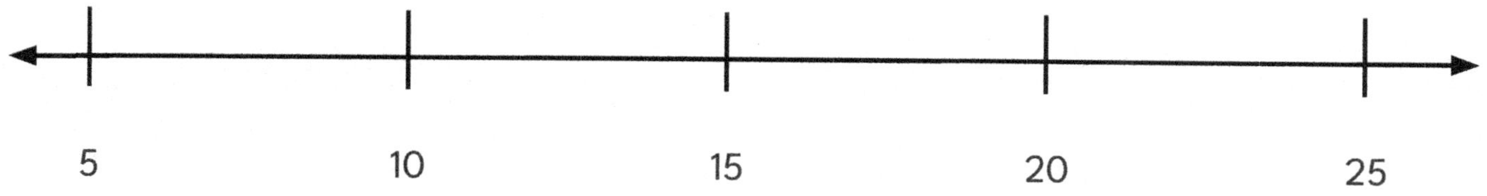

Explain how you found the missing tick marks.

2.MD.9

5) Draw and label tick marks that will show 15, 25, 35, and 45 on the number line.

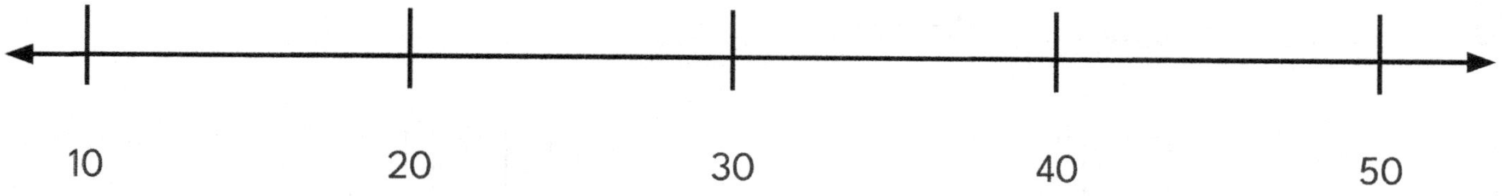

Explain how you found the missing tick marks.

2.MD.9

Do you understand? **?** ✓

Name: ______________________________ Date: ______________

Unit 4 Practice 4: I can compare numbers using a number line.

1) Write the missing number that makes each equation true.

15 - 10 = _______ 15 - _______ = 10

30 - 10 = _______ 30 - 5 = _______

45 - 15 = _______ 45 - _______ = 25

50 - 25 = _______ 50 - 35 = _______

40 - 15 = _______ 50 - _______ = 5

2.NBT.5

2) How many in each collection?

Use <, >, or = to compare the collections.

_______ ○ _______

1.NBT.3

Do you understand? **?** ✓

Name: ______________________________ Date: ______________

Unit 4 Practice 4: I can compare numbers using a number line.

3) Piper wants to compare the number 12 and the number 10 using a number line. Draw points on the number line that Piper will need to show the numbers 12 and 10.

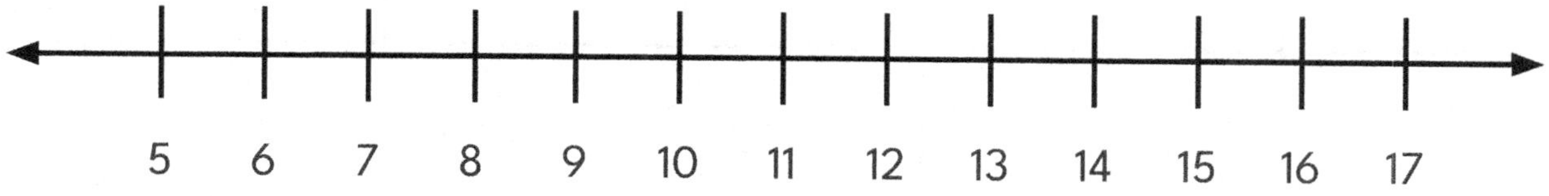

Piper then wrote 12 > 10.

Use what you know about the number line to help Piper explain her thinking.

2.MD.6

4) Draw points on the number line to show the numbers 4 and 7.

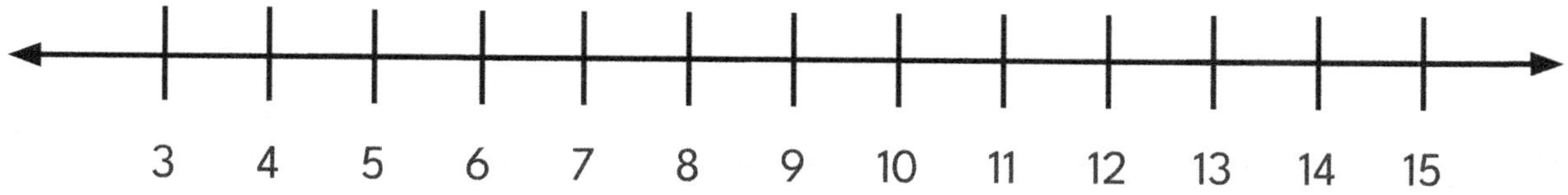

Use <, >, or = to compare the numbers 4 and 7.

4 ◯ 7

Use what you know about the number line to explain your thinking.

2.MD.6

Do you understand? **?** ✓

Name: ______________________________ Date: ______________

Unit 4 Practice 5: I can skip count and estimate number placement.

1) Skip count by five on the number line.

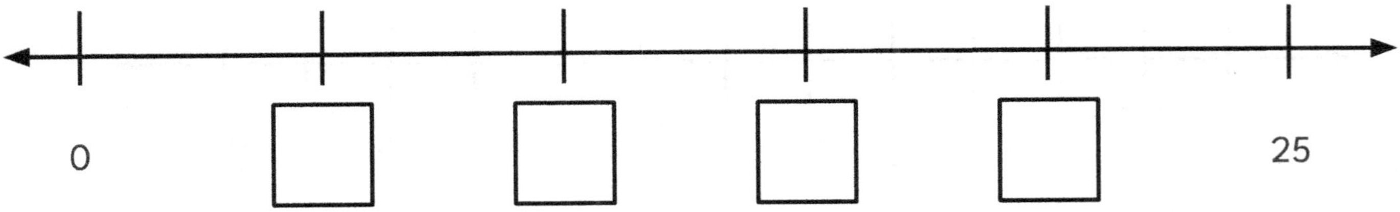

2.NBT.2

2) Skip count by ten on the number line.

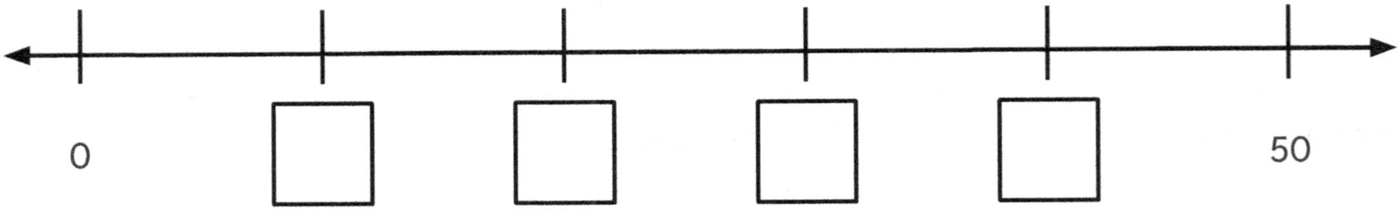

2.NBT.2

3) Fill in the missing numbers.

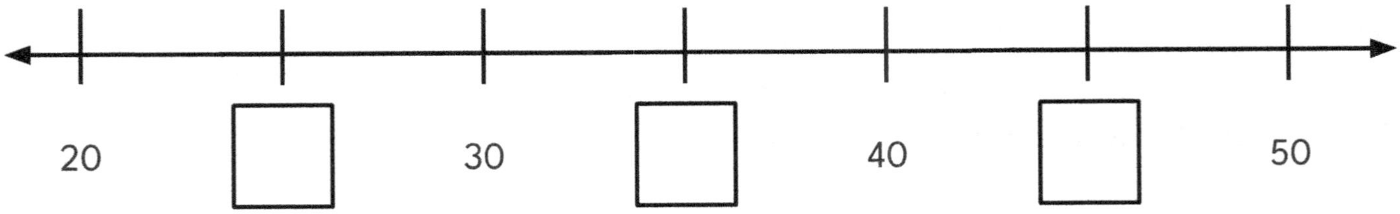

2.NBT.2

Do you understand? **?** ✓

Affirmation: Forgetting and making mistakes are a part of learning and remembering.

Name: ______________________________ Date: ______________

Unit 4 Practice 5: I can skip count and estimate number placement.

4) **Estimate** the number this point might show on the number line.

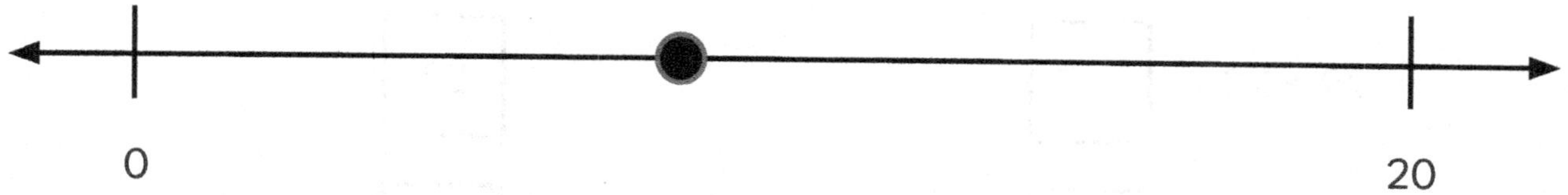

Use what you know about the number line to explain your thinking.

2.MD.6

5) **Estimate** the number this point might show on the number line.

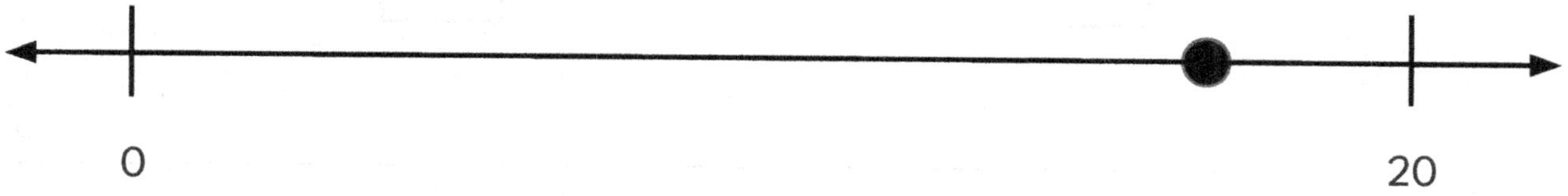

Use what you know about the number line to explain your thinking.

2.MD.6

6) **Estimate** the number this point might show on the number line.

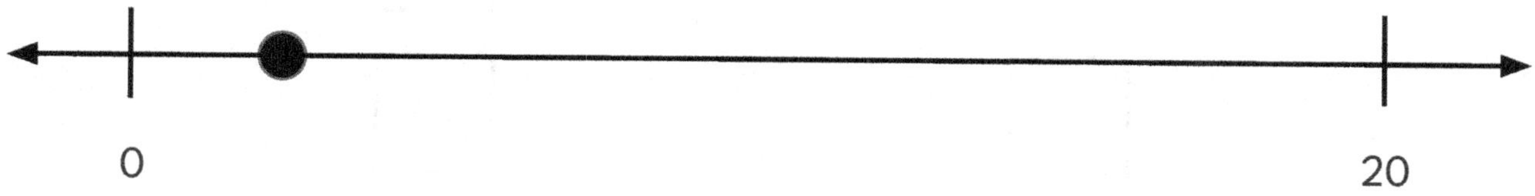

Use what you know about the number line to explain your thinking.

2.MD.6

Do you understand? **?** ✓

Name: ______________________________ Date: ______________

Unit 4 Practice 6: I can find numbers to 100 on a number line.

1) Match the number that shows the sum.

79 + 20 =	☐	28
14 + 14 =	☐	88
44 + 44 =	☐	70
19 + 19 =	☐	38

2.NBT.5

2) Match the number that shows the difference.

80 - 35 =	☐	15
29 - 14 =	☐	28
99 - 55 =	☐	45
47 - 19 =	☐	44

2.NBT.5

Do you understand? **?** ✓

Name: ______________________ Date: __________

Unit 4 Practice 6: I can find numbers to 100 on a number line.

3) Write numbers to compete the number line.

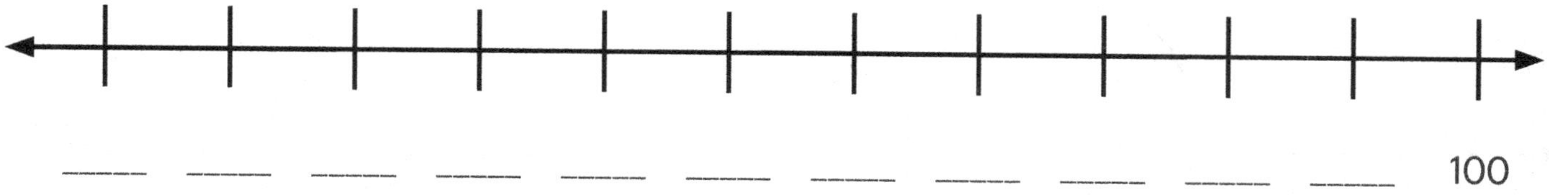

Write numbers to compete the number line in a different way.

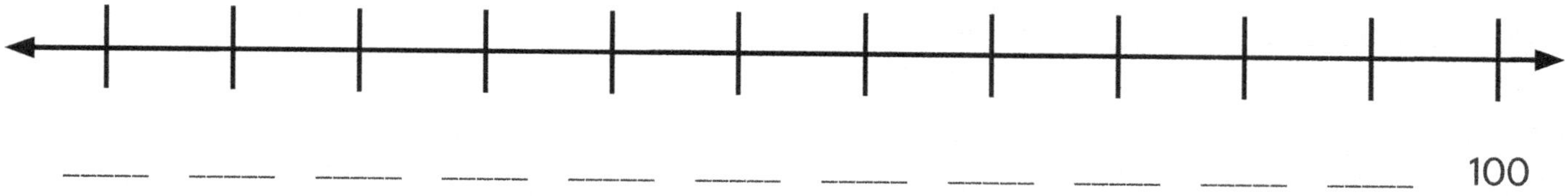

2.MD.6

4) Draw and label a point to estimate where the numbers 90, 99, 83, 91, and 89 might be found on the number line below.

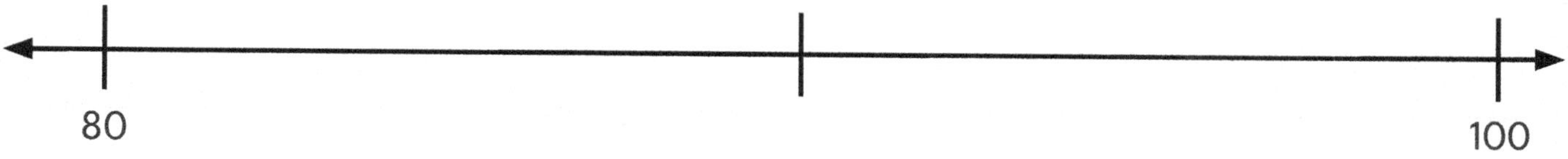

2.MD.6

Do you understand? **?** ✓

Name: ______________________________ Date: ______________

Unit 4 Practice 7: I can hop to the right or to the left on a number line.

1) Does the arrow show moving right or left on the number line?

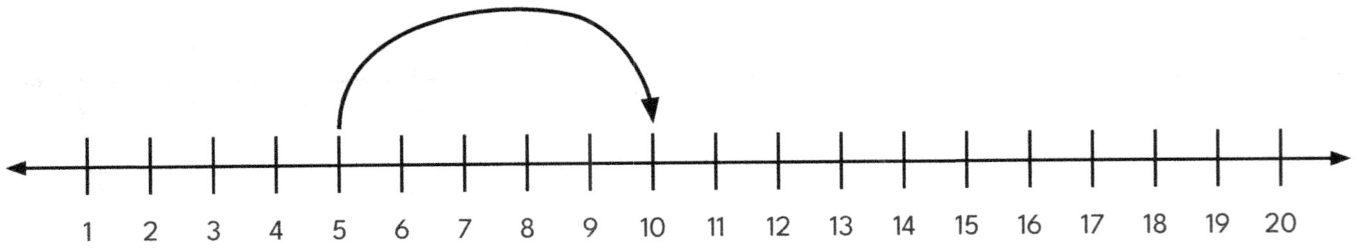

The arrow is showing that we are moving to the _______.

The arrow starts on ________ and jumps to _______.

How many hops did the arrow make?

The arrow makes _______ hops.

2.MD.6

2) Does the arrow show addition or subtraction?

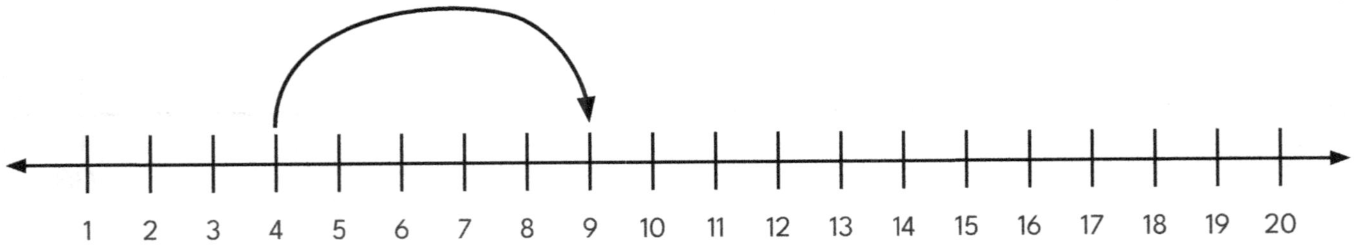

The arrow is moving to the right. This shows ____________________.

Circle the matching equation.

$4 + 5 = 9$ $9 - 5 = 4$

2.MD.6

Do you understand? **?** ✓

Affirmation: I am math. From the cells in my body to the hairs on my head. I am math.

Name: ______________________________ Date: ______________

Unit 4 Practice 7: I can hop to the right or to the left on a number line.

3) Does the arrow show moving right or left on the number line?

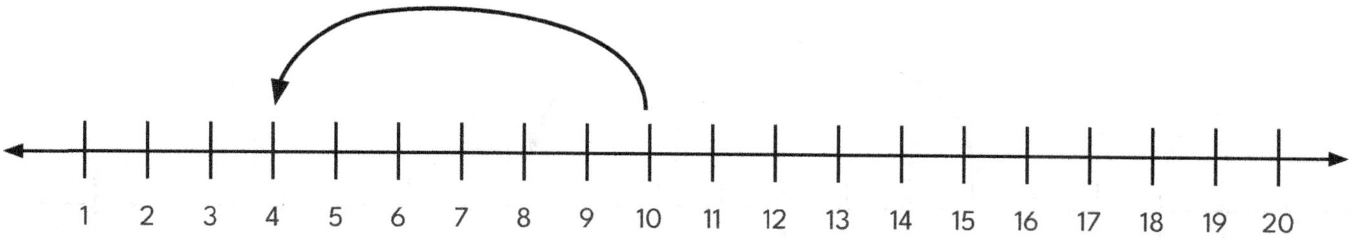

The arrow is showing that we are moving to the _______.

The arrow starts on ________ and jumps to _______.

How many hops did the arrow make?

The arrow makes _______ hops.

2.MD.6

4) Does the arrow show addition or subtraction?

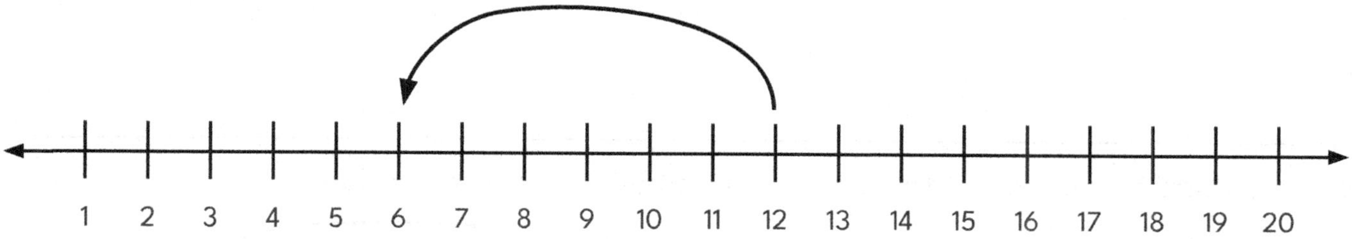

The arrow is moving to the left. This shows ____________________.

Circle the matching equation.

$4 + 5 = 9$ $9 - 5 = 4$

2.MD.6

Do you understand? **?** ✓

Affirmation: When I learn something new, I will forget some of it, and that is okay.

Name: ______________________________ Date: ______________

Unit 4 Practice 8: I can write an equation to match an arrow on a number line.

1) Write an addition equation to match the number line representation.

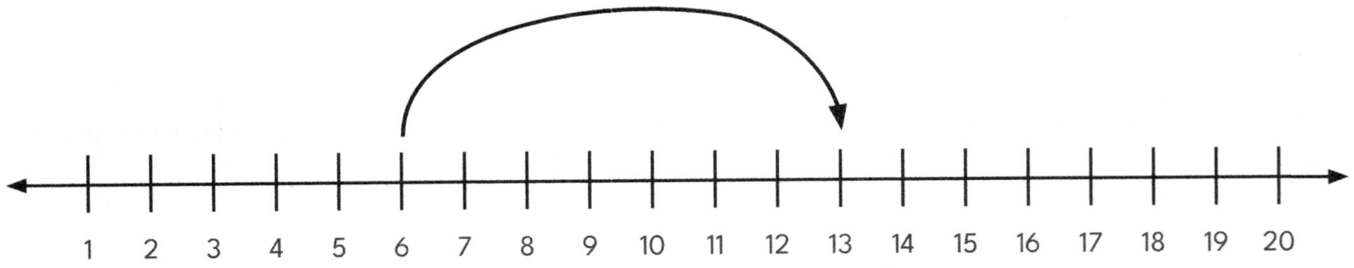

Equation: ______________________________

2.MD.6

2) Write an addition equation to match the number line representation.

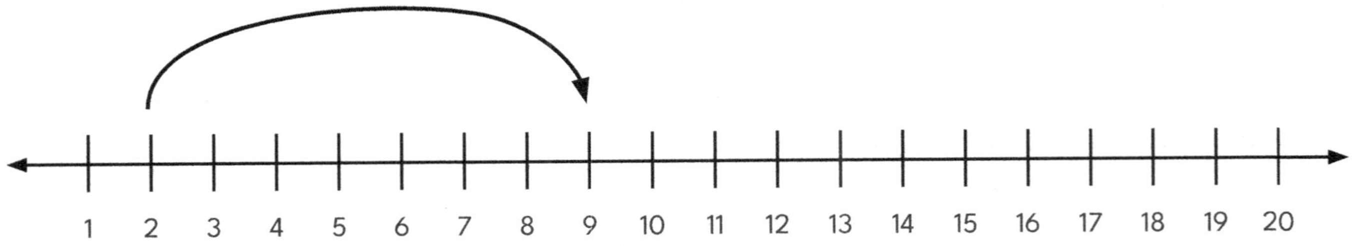

Equation: ______________________________

2.MD.6

3) Write an addition equation to match the number line representation.

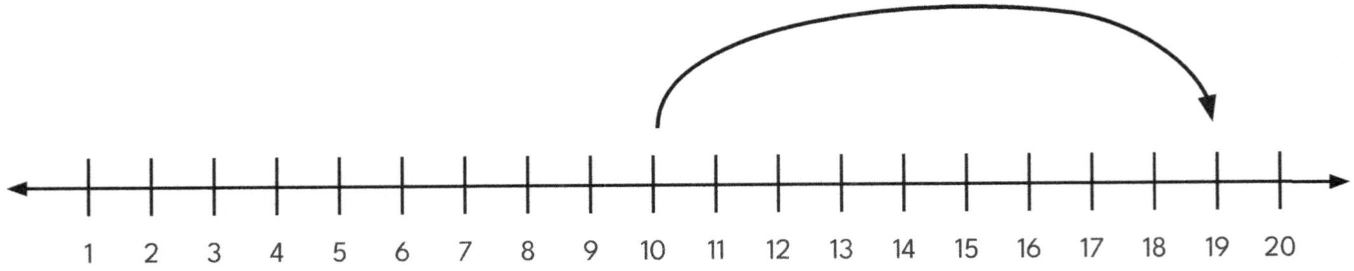

Equation: ______________________________

2.MD.6

Do you understand? **?** ✔

Name: ______________________ Date: ____________

Unit 4 Practice 8: I can write an equation to match an arrow on a number line.

4) Write a subtraction equation to match the number line representation.

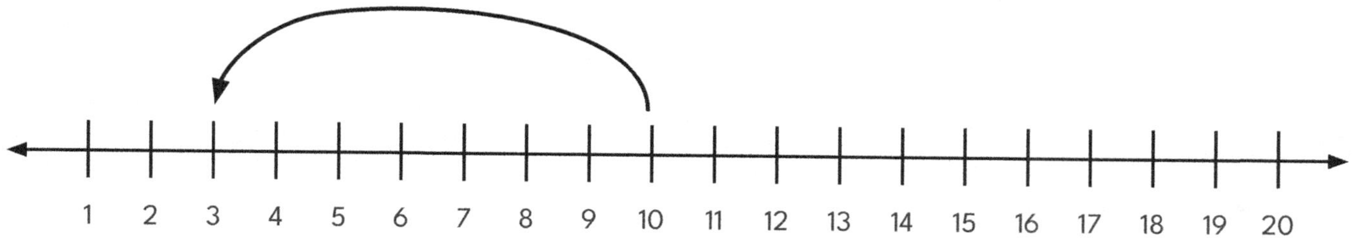

Equation: ______________________

2.MD.6

5) Write a subtraction equation to match the number line representation.

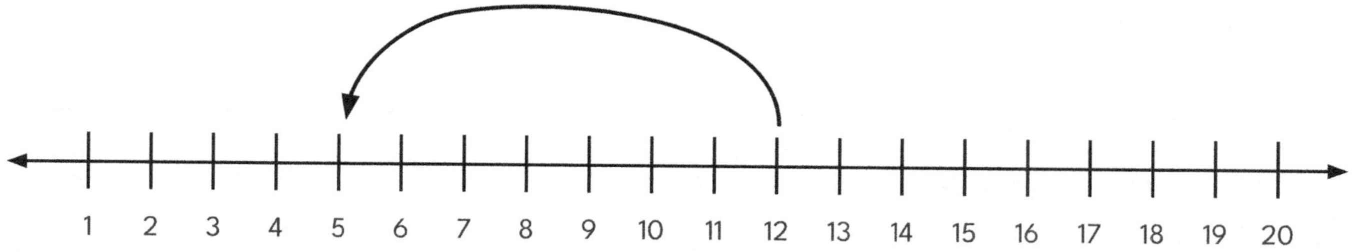

Equation: ______________________

2.MD.6

6) Write a subtraction equation to match the number line representation.

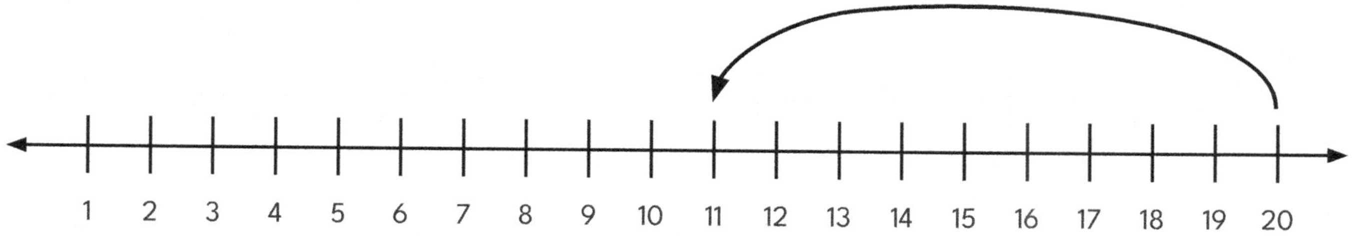

Equation: ______________________

2.MD.6

Do you understand? **?** ✓

Affirmation: I am a mathematical thinker.

Name: ______________________________ Date: ______________

Unit 4 Practice 9: I can model addition or subtraction on a number line.

1) Counting on and counting back.
Count on by ones.

12, _____, _____, _____, _____, _____, _____, _____, _____, 21

Count back by ones.

20, _____, _____, _____, _____, _____, _____, _____, _____, 11

1.NBT.1

2) Show 5 + 6 = 11 on the number line.

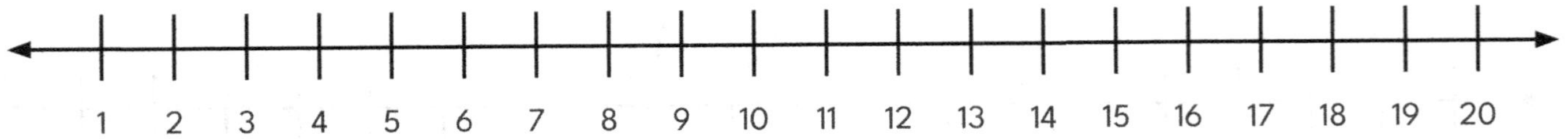

Does your arrow show you moving to the right or to the left?

My arrow is showing that I am moving to the ______________.

2.MD.6

3) Show 16 - 6 = 10 on the number line.

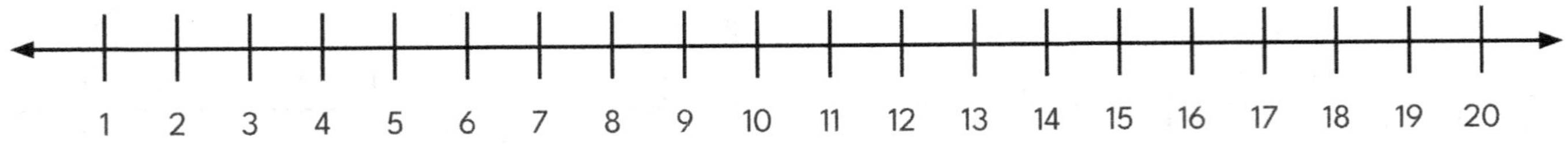

Does your arrow show you moving to the right or to the left?

My arrow is showing that I am moving to the ______________.

2.MD.6

Do you understand? **?** ✓

Name: ______________________________ Date: ______________

Unit 4 Practice 9: I can model addition or subtraction on a number line.

4) Show 7 + 2 = 9 on the number line.

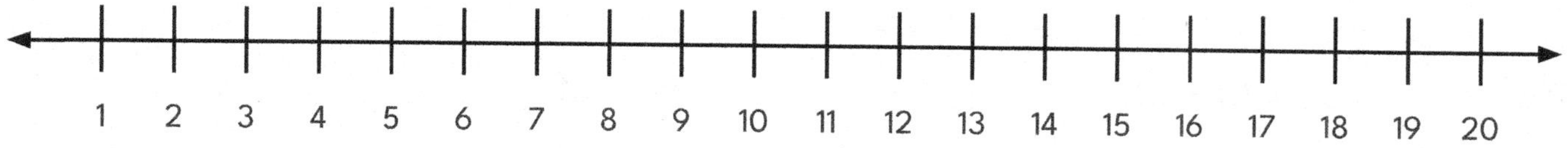

Does your arrow show you moving to the right or to the left?

My arrow is showing that I am moving to the ____________.

2.MD.6

5) Show 12 -4 = 8 on the number line.

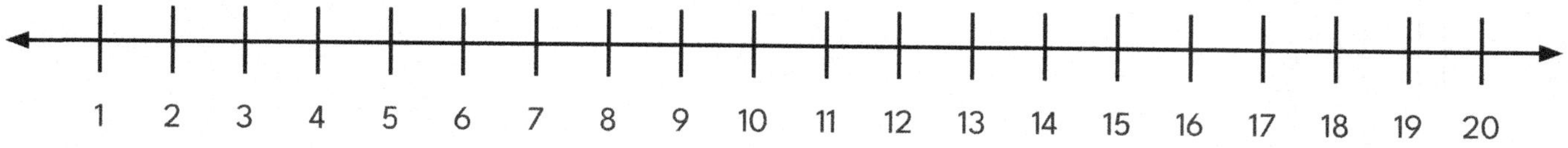

Does your arrow show you moving to the right or to the left?

My arrow is showing that I am moving to the ____________.

2.MD.6

6) Show 15 - 9 = 6 on the number line.

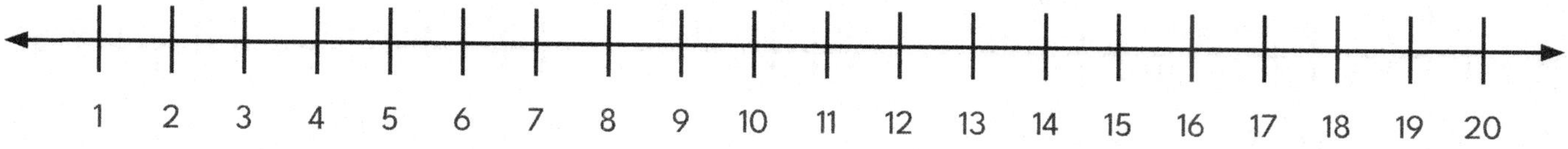

Does your arrow show you moving to the right or to the left?

My arrow is showing that I am moving to the ____________.

2.MD.6

Do you understand? **?** ✓

Name: ______________________________ Date: ______________

Unit 4 Practice 10: I can use base-ten diagrams and number lines to add.

1) Find the sum.

36 + 10 = ______ 56 + 20 = ______

36 + 10 + 10 = ______ 56 + 30 = ______

36 + 10 + 10 + 2 = ______ 56 + 30 + 4 = ______

1.NBT.4

2) Find the sum: 27 + 30
Finish drawing the base-ten diagram.

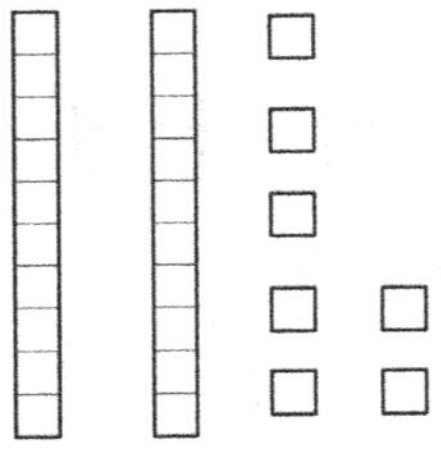

Equation: ____

1.NBT.4

3) Find the sum: 27 + 30

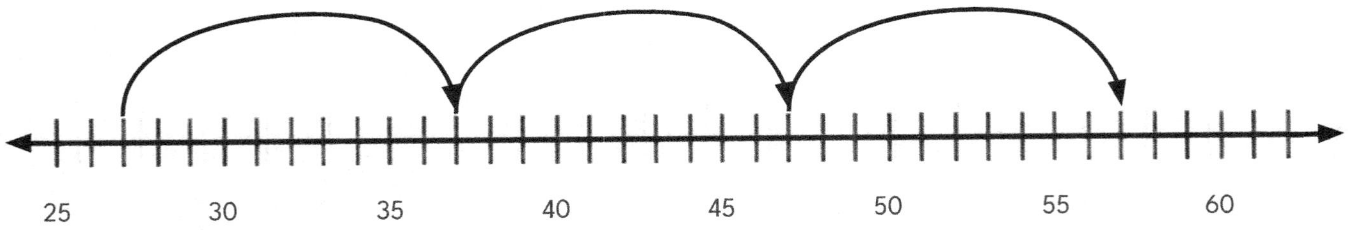

There are ______ jumps of ten.

Equation: ____

2.MD.6

Do you understand? **?** ✓

Affirmation: I am brave and I take risks in math.

Name: ______________________________ Date: ______________

Unit 4 Practice 10: I can use base-ten diagrams and number lines to add.

4) Solve 43 + 23 using a base-ten diagram.

Equation: ______________________________

2.NBT.5

5) Solve 43 + 23 using the number line.

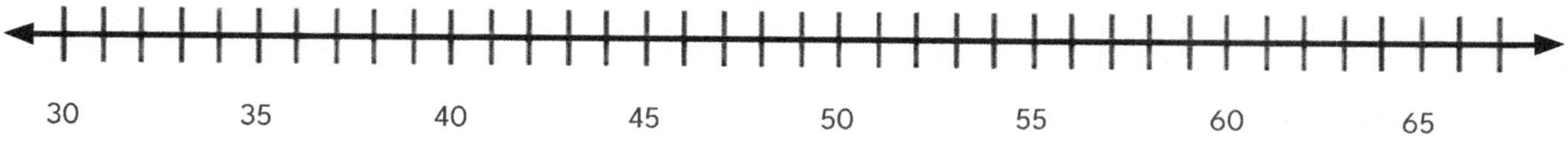

Equation: ______________________________

2.MD.6

6) Solve 34 + 25 using a base-ten diagram or the number line.

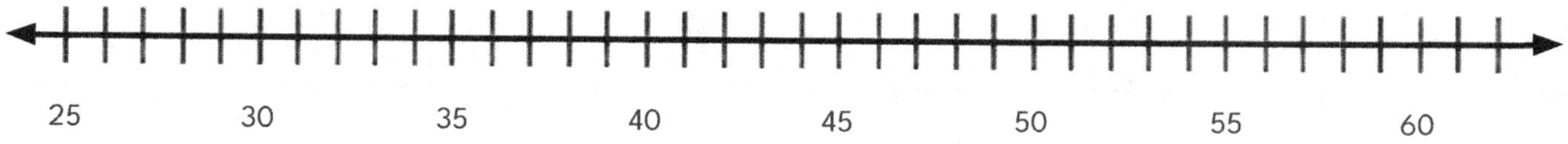

Equation: ______________________________

2.NBT.7

Do you understand? **?** ✓

Name: ______________________________ Date: ____________

Unit 4 Practice 11: I can use base-ten diagrams and number lines to subtract.

1) Find the difference.

46 - 10 = _______

37 - 20 = _______

46 - 10 - 10 = _______

37 - 30 = _______

46 - 10 - 10 - 2 = _______

78 - 30 - 4 = _______

1.NBT.6

2) Find the difference: 56 - 20

Use the base-ten representation to solve.

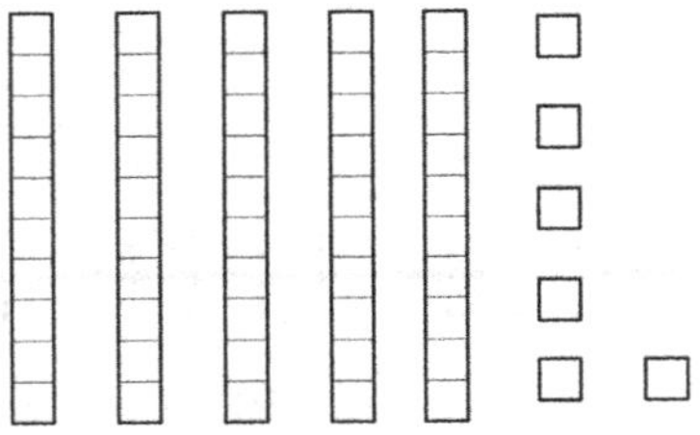

Equation: ______________________________

1.NBT.6

3) Find the difference: 56 - 20

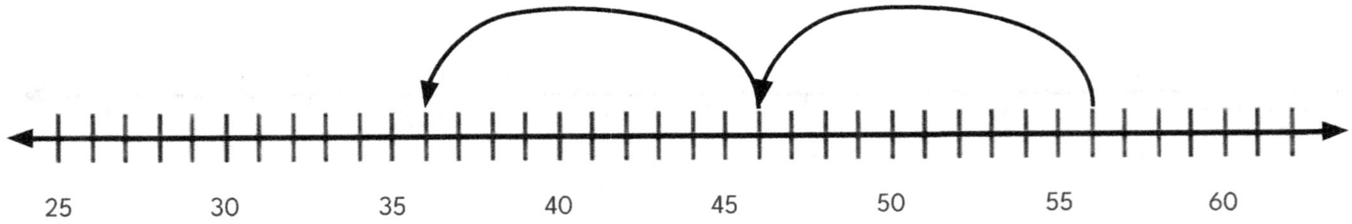

There are _______ jumps of ten.

Equation: ______________________________

2.MD.6

Do you understand? ? ✓

Affirmation: Math is more than just getting an answer.

Name: ______________________________ Date: ______________

Unit 4 Practice 11: I can use base-ten diagrams and number lines to subtract.

4) Solve 64 - 21 using base-ten representations.

Equation: ______________________________

2.NBT.5

5) Solve 64 - 21 using the number line.

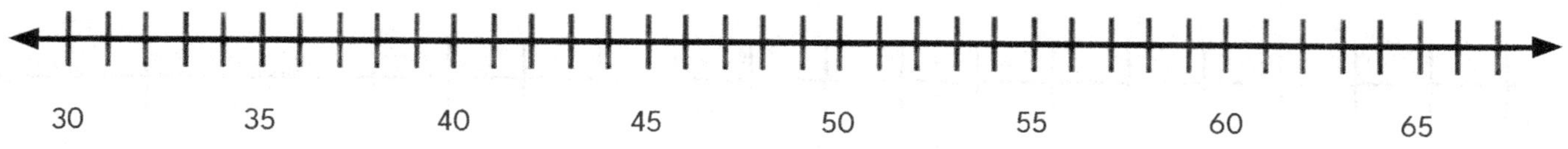

Equation: ______________________________

2.MD.6

6) Solve 59 - 14 using base-ten representations or a number line.

Equation: ______________________________

2.NBT.7

Do you understand? **?** ✓

Name: ______________________________ Date: ______________

Unit 4 Practice 12: I can use a number line to write an equation.

1) **Sample:** Steps to show an addition equation on a number line.

Start number: 9 Jump 5 to the right. End number: 14

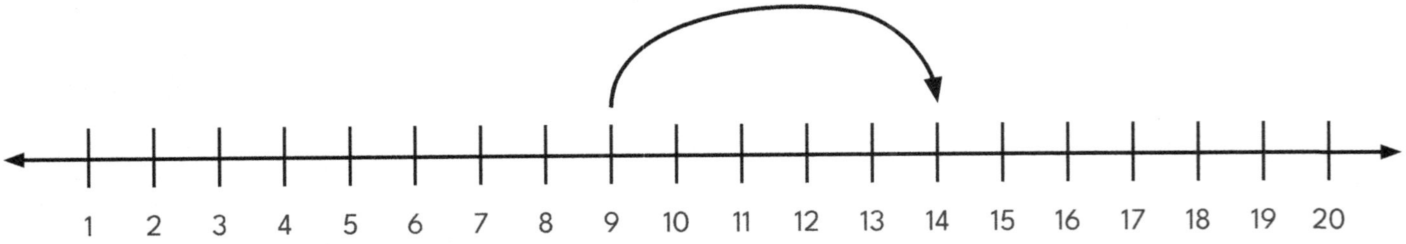

Equation: 9 + 5 = 14

2.MD.6

2) Follow the steps to show the addition equation on the number line.

Start number: 4 Jump 8 to the right. End number: 12

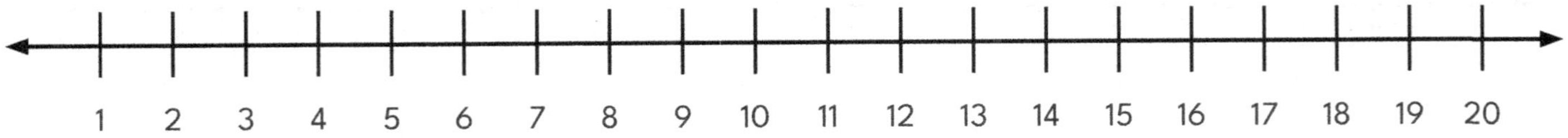

Equation: 4 + 8 = 12

2.MD.6

3) Make your own.

Start number: _______ Jump _______ to the _______ End number: _______

Equation: ______________________________

2.MD.6

Do you understand? **?** ✓

Name: ______________________________ Date: ______________

Unit 4 Practice 12: I can use a number line to write an equation.

4) **Sample:** Steps to show a subtraction equation on a number line.

Start number: 11 Show a jump of 6 to the left. End number: 5

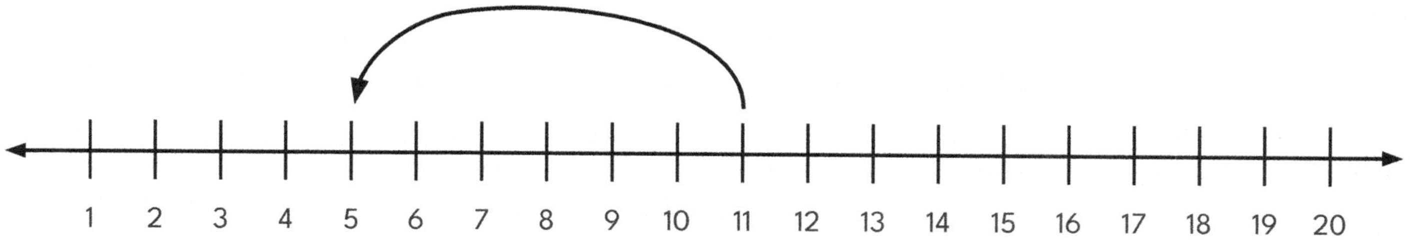

Equation: 11 - 6 = 5

2.MD.6

5) Follow the steps to show the subtraction equation on the number line.

Start number: 14 Show a jump 5 to the left. End number: 9

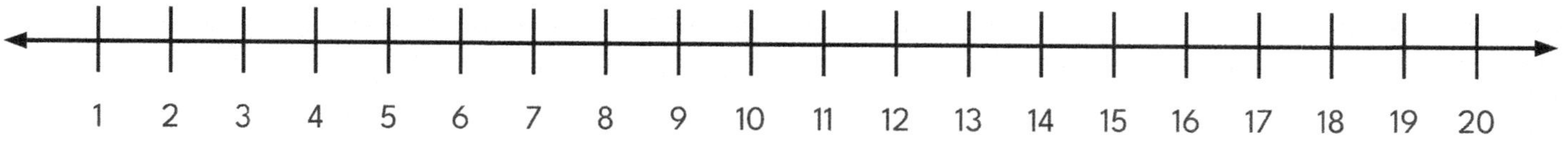

Equation: 14 - 5 = 9

2.MD.6

6) Make your own.

Start number: _______ Jump _______ to the _______ End number: _______

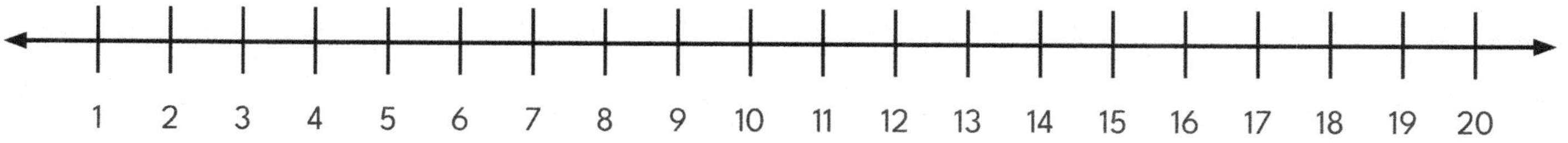

Equation: ______________________________

2.MD.6

Do you understand? **?** ✓

Affirmation: It is important that I explain and justify my thinking.

Name: ______________________________ Date: ______________

Unit 4 Practice 13: I can solve story problems using number lines or diagrams.

1) Use the tape diagram to show 35 + ? = 50 on the number line.

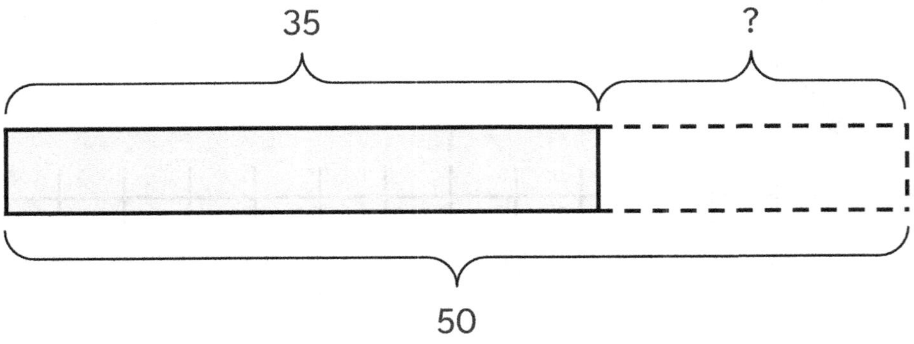

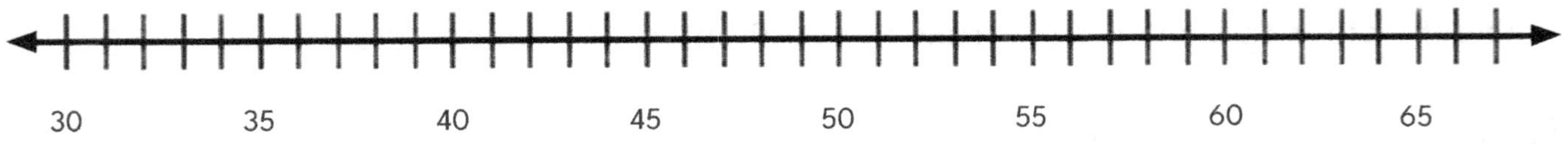

Equation: ______________________________

1.OA.4

2) Abby has 34 blue stickers and 30 green stickers. How many stickers does she have? Use the tape diagram and the number line to show your thinking.

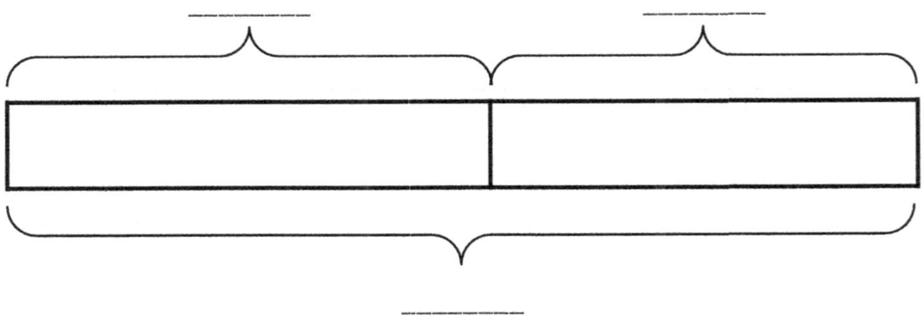

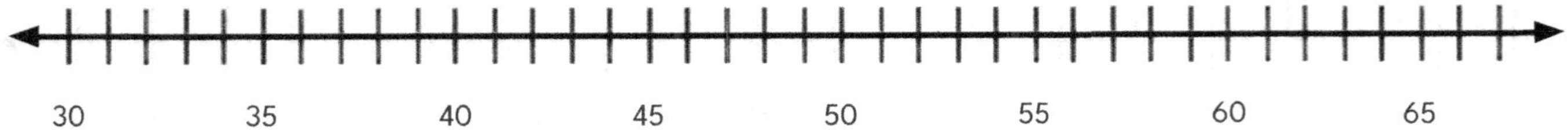

Equation: ______________________________

2.OA.1

Do you understand? **?** ✓

Affirmation: It is important that I explain and justify my thinking.

Name: ______________________________ Date: ______________

Unit 4 Practice 13: I can solve story problems using number lines or diagrams.

3) Mason has 19 shapes. Some are triangles and 12 are squares. How many triangles does he have?

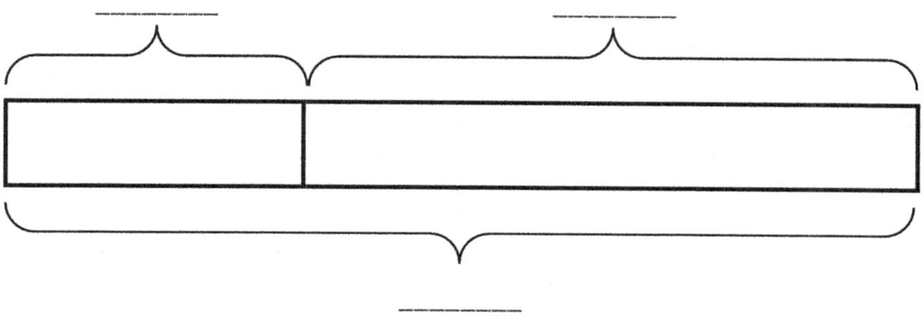

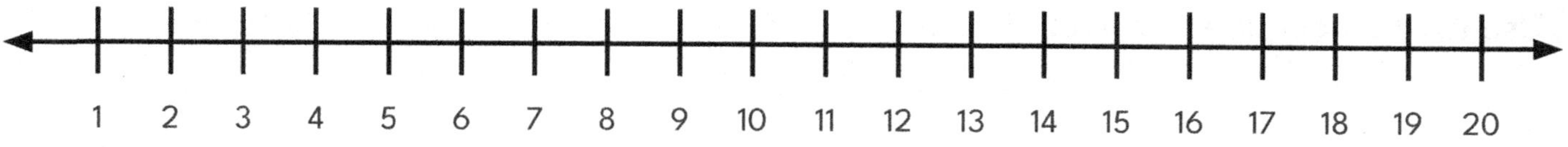

Equation: ______________________________

2.OA.1

4) A pet store has 32 goldfish and some tetras for sale in a tank. The are 65 fish in the tank. How many tetras are in the fish tank?

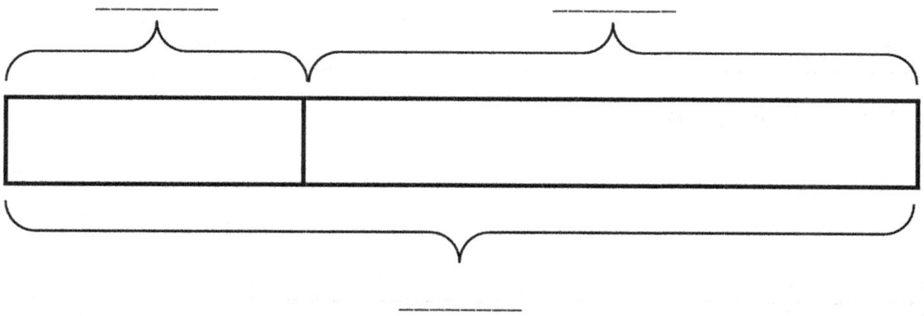

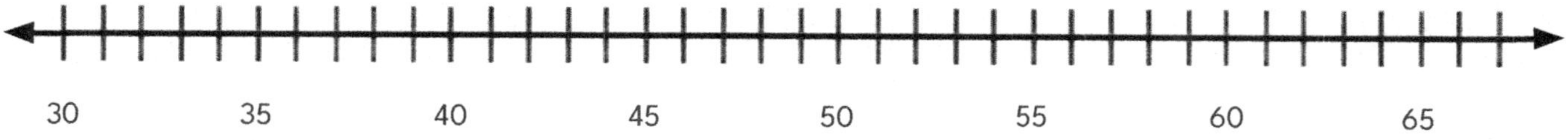

Equation: ______________________________

2.OA.1

Do you understand? **?** ✓

Name: ______________________________ Date: ______________

Unit 4 Practice 14: I can use a number line to add or subtract.

1) Count on by tens.

28, _____, _____, _____, _____, _____, _____, _____, 108

Count on by ones.

67, _____, _____, _____, _____, _____, _____, _____, _____, 76

1.2.NBT.2

2) Use the number line to solve: 68 + 24

Equation: ______________________________

2.MD.6

3) Use the number line to solve: 14 + 27

Equation: ______________________________

2.MD.6

Do you understand? ? ✓

Affirmation: We are all smart in different ways.

Name: ______________________________ Date: ______________

Unit 4 Practice 14: I can use a number line to add or subtract.

4) Count back by tens.

98, ______, ______, ______, ______, ______, ______, ______, ______, 8

Count back by ones.

65, ______, ______, ______, ______, ______, ______, ______, ______, 56

2.NBT.2

5) Use the number line to solve: 68 - 24

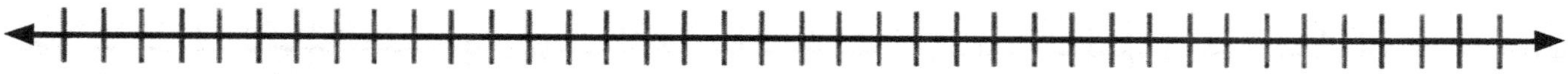

Equation: ______________________________

2.MD.6

6) Use the number line to solve: 27 - 14

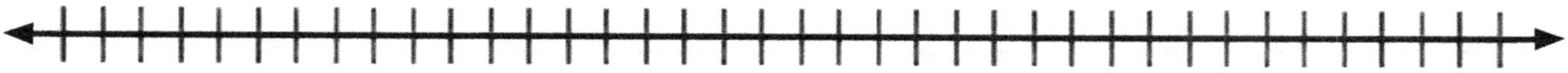

Equation: ______________________________

2.MD.6

Do you understand? **?** ✓

Name: ______________________________ Date: ____________

Unit 4 Practice 15: I can solve or write a number riddle using a number line.

1) Count on by fives.

22, ______, ______, ______, ______, ______, ______, ______, ______, 67

Count back by fives.

49, ______, ______, ______, ______, ______, ______, _____, ______, 4

2.NBT.2

2) Use the number line to solve the riddle.
Peyton started at 55 and jumped 26 to the right. Where did Peyton end?

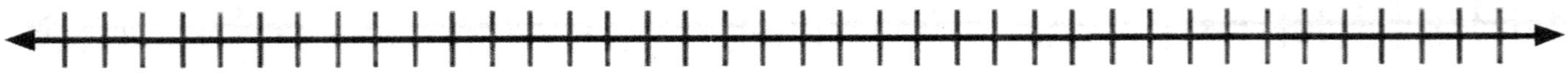

Equation: ____

2.MD.6

3) Use the number line to solve the riddle.
Cory started at 98 and jumped 26 to the left. Where did Cory end?

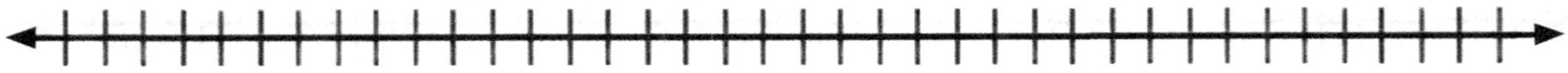

Equation: ____

2.MD.6

Do you understand? **?** ✓

Name: ______________________________ Date: ______________

Unit 4 Practice 15: I can solve or write a number riddle using a number line.

4) Use the information on the number line.

Start number: _______ Jump of _______ to the __________. End number: _______

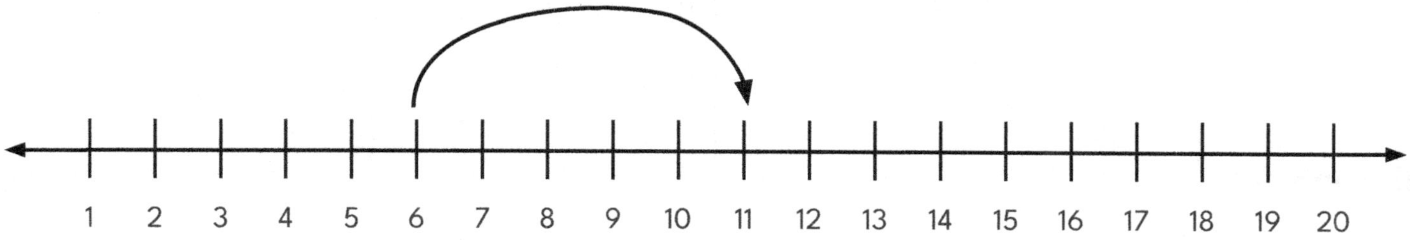

Equation: ______________________________

2.MD.6

5) Write a number riddle.

Start number: _______ Jump of _______ to the __________. End number: _______

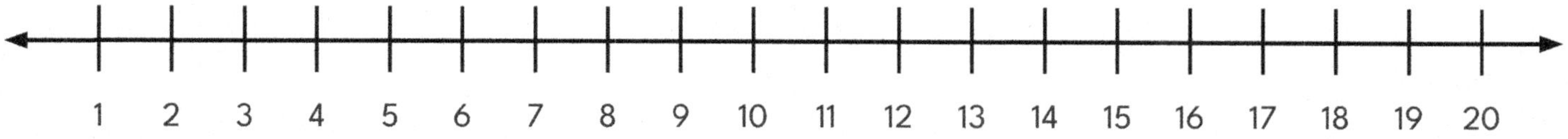

Equation: ______________________________

2.MD.6

6) Write a number riddle.

Start number: _______ Jump of _______ to the __________. End number: _______

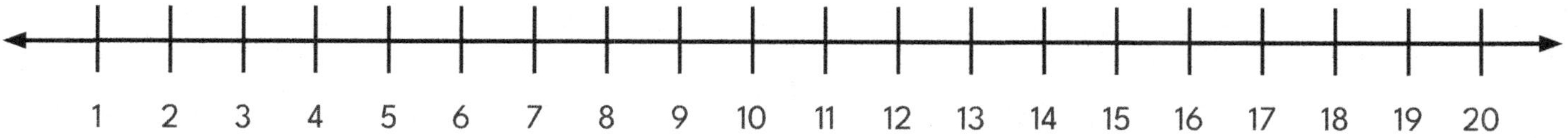

Equation: ______________________________

2.MD.6

Do you understand? **?** ✓

Name: ______________________________ Date: ______________

I can practice grade level fluencies.

Set 1: Add five	**Set 2:** Subtract five
1) 14 + 5 = _______	1) 19 - 5 = _______
2) 15 + 5 = _______	2) 20 - 5 = _______
3) 60 + 5 = _______	3) 50 - 5 = _______
4) 65 + 5 = _______	4) 80 - 5 = _______
5) 24 + 5 = _______	5) 26 - 5 = _______
6) 25 + 5 = _______	6) 25 - 5 = _______
7) 26 + 5 = _______	7) 24 - 5 = _______
2.NBT.2	2.NBT.2

Do you understand? **?** ✓

Unit Reflection

Addition and Subtraction on the Number Line

Use this space to reflect on your understanding of the unit skills and concepts.

Skill/Concept	I can...	I need to work on...

Made in the USA
Monee, IL
08 November 2026

67383087R00096